# Lecture Notes in Mathematics

Edited by A. Dold and B. Eckmann

805

# Oldřich Kowalski

# Generalized Symmetric Spaces

Springer-Verlag
Berlin Heidelberg New York 1980

**Author**
Oldřich Kowalski
Department of Mathematical Analysis
Faculty of Mathematics and Physics
Charles University, Sokolovská 83
18600 Prague/Czechoslovakia

AMS Subject Classifications (1980): 15A21, 17B40, 22E25, 53C05, 53C20, 53C30, 53C35, 53C55

ISBN 3-540-10002-4 Springer-Verlag Berlin Heidelberg New York
ISBN 0-387-10002-4 Springer-Verlag New York Heidelberg Berlin



Printing and binding: Beltz Offsetdruck, Hemsbach/Bergstr.
2141/3140-543210

# P R E F A C E

In this booklet we present an elementary theory of generalized symmetric spaces. "Elementary" means here that topological invariants and the advanced parts of algebra (like root systems) are not involved. There are also more sophisticated theories that simulate into a great depth the theory of ordinary symmetric spaces ([WoG],[Gr]) - yet in these theories one needs to put substantial restrictions on the basic group of the space (e.g. semi-simplicity). We refer briefly to some of these advanced results in Note 5. In our exposition we do not put any such restrictions on groups.

We try to develop our theory as a part of the theory of Riemannian spaces or spaces with an affine connection rather than a part of the theory of Lie groups and Lie algebras. (For a group-theoretical direction see the booklet [F3].) So far, the expression "generalized symmetric space" has been usually used only as a nickname denoting various geometrical or group-theoretical concepts (usually a triplet $(G,H,\sigma)$, where $G$ is a Lie group, $H$ a closed Lie subgroup of $G$ and $\sigma$ an automorphism of $G$ which is not necessarily involutive). The present author tries to give a precise and purely geometrical meaning to this expression; the basic new concepts are "generalized symmetric Riemannian space" and "generalized affine symmetric space". (In the same spirit, we replace the group-theoretical concept of a reductive homogeneous space by the more geometrical concept of an affine reductive space.)

The book does not contain the theory of symmetric spaces as a special part (why duplicate S.Helgason or O.Loos?) and also our references to symmetric spaces are only exceptional (this is not the right place to pay a tribute to E.Cartan and to other great men of the past). A good deal of the book is devoted to the questions which have no analogue in the theory of the ordinary symmetric spaces; in particular, to the detailed study of eigenvalues of generalized symmetries.

The theory of generalized symmetric pseudo-Riemannian spaces is also not included, in spite of the fact that a considerable progress has been made in the theory of pseudo-Riemannian symmetric spaces ([B], [CP1],[CP2],[CW]) which may stimulate the interest in the generalizations, too. (See also Note 5.)

Further, the book is not a survey of what has been done in generalizations of symmetric spaces; it is only a lecture note with a restricted subject. It was inspired primarily by works of A.J.Ledger, O.Loos and others. As far as I know, the only survey article has been published by V.I.Vedernikov and A.S.Fedenko in 1976 (see [VF]).

The list of references is not intended to be a collection of prestigeous papers and famous names. The references are simply limited to those closely connected with the topic. Thus some works quoted here are important and some others may be not.

Now, let us sum up briefly the contents:

In Chapter 0 we simply try to catch the reader's eye. It contains basic information on generalized symmetric Riemannian spaces.

Chapter 1 is a short exposé of the theory of reductive homogeneous spaces. The style differs a bit from the traditional one. The paragraph on affine reductive spaces might be of interest by itself.

Chapter 2 deals with a kind of differentiable distributive groupoids, so-called regular s-manifolds. We generalize here the abstract symmetric spaces introduced by O.Loos. The canonical connection, automorphisms and transvections are studied. We also prove that regular s-manifolds are reductive homogeneous spaces.

Chapter 3 generalizes the theory of locally symmetric spaces. We also introduce infinitesimal models of regular s-manifolds, characterizing the latter up to a local automorphism. (Alternatively, we can characterize regular s-manifolds by "generalized symmetric" Lie algebras.) In the last paragraph we study the invariant almost complex structures on locally regular s-manifolds.

In Chapter 4 we study invariant submanifolds and various operations with regular s-manifolds; some of them have no analogue in the theory of ordinary symmetric spaces.

Chapter 5 is devoted to the following topic: in an ordinary symmetric space we always have <u>a canonical</u> family of symmetries. In a generalized Riemannian or affine symmetric space we usually have many <u>admissible</u> families of (generalized) symmetries. None of them is "canonical" but some of them are said to be "distinguished". If we represent the generalized symmetric spaces by their distinguished families of symmetries, we can get a better insight into the classification problems. For this purpose, it is necessary to study the relations among the eigenvalues of generalized symmetries, which is an interesting combinatorial problem.

In Chapter 6 we present the local classification of generalized symmetric Riemannian spaces of dimension $n \leq 5$. The complete proofs are given only for dimensions 3 and 4. We limit ourselves to the spaces which are <u>not</u> Riemannian symmetric - the latter are known from the famous list of E.Cartan.

In Chapter 7 we present the local classification of generalized affine symmetric spaces of dimensions 3 and 4. The proofs (including long and cumbersome calculations) are omitted. Again, the classification does not involve ordinary affine symmetric spaces. (Contrary to the Riemannian case, the complete classification of affine symmetric spaces is known only for the spaces with a semi-simple basic group, see [B].)

The book assumes a basic knowledge of the modern differential geometry (cf. Foundations of Differential Geometry by S.Kobayashi and K.Nomizu, Vol.I), and of the theory of Lie groups and Lie algebras (cf. C.Chevalley, Theory of Lie Groups I). The reader can also consult Appendix A and B.

<u>Acknowledgement</u>. The author wishes to thank to A.Gray for his valuable hints and comments during the preparation of this lecture note, and to P.Dombrowski for his great encouragement and help.

Prague, September 1978 Oldřich Kowalski

# TABLE OF CONTENTS


LIST OF STANDARD DENOTATIONS . . . . . . . . . . . . . . . . . . . . . XI

CHAPTER 0 - GENERALIZED SYMMETRIC RIEMANNIAN SPACES

Riemannian symmetric spaces . . . . . . . . . . . . . . . 1
Riemannian s-structures . . . . . . . . . . . . . . . . . 2
Regular s-structures . . . . . . . . . . . . . . . . . . 6
Existence theorems . . . . . . . . . . . . . . . . . . . 9
A low-dimensional example . . . . . . . . . . . . . . . 18
The de Rham decomposition . . . . . . . . . . . . . . . 20
Parallel and non-parallel s-structures . . . . . . . . 22
The canonical connection . . . . . . . . . . . . . . . 23

CHAPTER I - REDUCTIVE SPACES

Reductive homogeneous spaces . . . . . . . . . . . . . 27
The canonical connection . . . . . . . . . . . . . . . 28
Algebraic characterization . . . . . . . . . . . . . . 33
The group of transvections . . . . . . . . . . . . . . 36
Affine reductive spaces . . . . . . . . . . . . . . . . 41

CHAPTER II - DIFFERENTIABLE s-MANIFOLDS

Affine symmetric spaces . . . . . . . . . . . . . . . . 45
The main theorem . . . . . . . . . . . . . . . . . . . 46
The canonical connection . . . . . . . . . . . . . . . 48
Regular homogeneous s-manifolds . . . . . . . . . . . . 52
The group of transvections . . . . . . . . . . . . . . 57
s-manifolds of finite order . . . . . . . . . . . . . . 62
Riemannian regular s-structures . . . . . . . . . . . . 63
Metrizable regular s-manifolds . . . . . . . . . . . . 65
Disconnected regular s-manifolds . . . . . . . . . . . 66

CHAPTER III - LOCALLY REGULAR s-MANIFOLDS

Localization of the previous theory . . . . . . . . . . . 68
Infinitesimal s-manifolds . . . . . . . . . . . . . . . 73
Local regular s-triplets . . . . . . . . . . . . . . . 76
A construction of local regular s-structures . . . . . . 78
The Riemannian case . . . . . . . . . . . . . . . . . . 81
Invariant almost complex structures . . . . . . . . . . . 84

CHAPTER IV - OPERATIONS WITH s-MANIFOLDS

Submanifolds and foliations . . . . . . . . . . . . . . 89
Decomposition of s-manifolds . . . . . . . . . . . . . . 93
Periodic s-manifolds . . . . . . . . . . . . . . . . . 96
Amalgamations . . . . . . . . . . . . . . . . . . . . . 99
Complexifications . . . . . . . . . . . . . . . . . . . 102
Spaces without infinitesimal rotations . . . . . . . . . 104
Pseudo-duality . . . . . . . . . . . . . . . . . . . . 106
Properties of eigenvalues . . . . . . . . . . . . . . . 107

CHAPTER V - DISTINGUISHED s-STRUCTURES ON GENERALIZED SYMMETRIC SPACES

Generalized affine symmetric spaces . . . . . . . . . . . 111
The multiplicative theory of eigenvalues . . . . . . . . 116
The additive theory of eigenvalues . . . . . . . . . . . 120
Applications and problems . . . . . . . . . . . . . . . 127

CHAPTER VI - THE CLASSIFICATION OF GENERALIZED SYMMETRIC RIEMANNIAN SPACES IN LOW DIMENSIONS

The classification procedure . . . . . . . . . . . . . . 131
Dimension n = 3 . . . . . . . . . . . . . . . . . . . . 134
Dimension n = 4 . . . . . . . . . . . . . . . . . . . . 136
Dimension n = 5 . . . . . . . . . . . . . . . . . . . . 142
Dimension n = 6 . . . . . . . . . . . . . . . . . . . . 148

CHAPTER VII - THE CLASSIFICATION OF GENERALIZED AFFINE SYMMETRIC SPACES IN LOW DIMENSIONS . . . . . . . 149

Note 1 - Existence of generalized symmetric spaces of solvable type . . . . . . . . . . . . . . . . . 158

Note 2 - Irreducible generalized affine symmetric spaces . . . . . . . . . . . . . . . . . . . . . . 161

Note 3 - Generalized pointwise symmetric spaces . . . . . . . 164

Note 4 - Non-parallel s-structures on symmetric spaces . . . . 168

Note 5 - Some advanced results . . . . . . . . . . . . . . . . 171

Appendix A: A digest of the theory of connections . . . . . . 173

Appendix B: Some theorems from the differential geometry . . 177

REFERENCES . . . . . . . . . . . . . . . . . . . . . . . . . 179

SUBJECT INDEX . . . . . . . . . . . . . . . . . . . . . . . 182

NOTATION INDEX . . . . . . . . . . . . . . . . . . . . . . . 187

# LIST OF STANDARD DENOTATIONS

| | |
|---|---|
| $M$ | a differentiable manifold of dimension $n$ and of class $C^\infty$ ("smooth manifold") |
| $T_p(M)$, $M_p$ | the tangent space of $M$ at the point $p \in M$ |
| $T_p^*(M)$, $M_p^*$ | the cotangent space of $M$ at $p$ |
| $\mathcal{T}(V)$ | the tensor algebra over a vector space $V$ |
| $\varphi_*$, $\varphi_{*p}$ | the tangent map of a smooth map $\varphi$ (on a manifold, at a point) |
| $T(M)$ | the tangent bundle of $M$ |
| $L(M)$ | the principal frame bundle of $M$ |
| $\mathcal{F}(M)$ | the ring of all smooth functions on $M$ |
| $\mathfrak{X}(M)$ | the Lie algebra of all smooth vector fields on $M$ |
| $\mathcal{T}(M)$ | the tensor algebra of all smooth tensor fields on $M$ |
| $G^\bullet$ | the identity component of a Lie group $G$ |
| $K_o$ | the isotropy subgroup of a transformation group $K$ (acting on $M$) at the point $o \in M$ |
| $R_x$, $L_x$ | the right and left translations of a Lie group |
| exp | the exponential map at the identity of a Lie group $G$ (considered as a map of the Lie algebra $\underline{g}$ into $G$) |
| $Ad_G$, $Ad$ | the representation of $G$ by the inner automorphisms of $G$ |
| $ad_G$, $(ad_{\underline{g}})$ | the adjoint representation of a Lie group $G$ (or of a Lie algebra $\underline{g}$) in the Lie algebra $\underline{g}$ of $G$ |
| ad | abbreviation for $ad_G$, $ad_{\underline{g}}$ |
| $\mathcal{L}_X$ | the Lie derivative with respect to $X \in \mathfrak{X}(M)$ |
| $[\ ,\ ]$ | the Lie bracket of a Lie algebra |
| $[\underline{a}, \underline{b}]$ | the vector subspace generated by all elements of the form $[X,Y]$, $X \in \underline{a}$, $Y \in \underline{b}$, where $\underline{a}$, $\underline{b}$ are vector subspaces of a Lie algebra |

(continued)

(continued)

| | |
|---|---|
| $\nabla$, $\nabla_X$ | affine connection on $M$, covariant derivative with respect to $X \in \mathfrak{X}(M)$ |
| $\mathrm{Exp}_p$ | the exponential map at a point of a manifold with the affine connection |
| $P(u_o)$ | the holonomy subbundle (of a connection) containing the frame $u_o \in L(M)$ |
| $\Phi(u_o)$ | the holonomy group with reference frame $u_o$ |
| $\Psi(p)$ | the holonomy group with reference point p |
| $\Phi^\bullet(u_o)$, $\Psi^\bullet(p)$ | the restricted holonomy groups |
| $I(M,g)$, $I(M)$ | the full group of isometries of a Riemann manifold |
| $A(M,\nabla)$, $A(M)$ | the full group of affine transformations of a manifold with the affine connection |

GENERAL CONVENTIONS:

All manifolds, maps, tensor fields etc. are supposed to be of class $C^\infty$ if not otherwise stated.

We use the symbol □ for the end of a proof and the symbol □□ for the end of an interrupted proof.

The star * indicates the sections or results which are published here for the first time (and no reference is available).

# CHAPTER 0

# GENERALIZED SYMMETRIC RIEMANNIAN SPACES

## Riemannian symmetric spaces.

In this paragraph we shall recall some elementary properties of the Riemannian symmetric spaces.

Let $(M,g)$ be a Riemannian manifold and $\nabla$ its Riemannian connection. Let $p \in M$ be a fixed point and $N_o$ a symmetric normal neighbourhood of the origin in the tangent space $M_p$ (i.e., for $X \in N_o$ we always have $-X \in N_o$). Put $N_p = \mathrm{Exp}_p(N_o)$, then $N_p$ is a "symmetric" normal neighbourhood of $p$.

For each point $x \in N_p$ consider the geodesic $t \mapsto \gamma(t)$ in $N_p$ such that $\gamma(0) = p$, $\gamma(1) = x$. Put $x' = \gamma(-1)$. Then the map $x \mapsto x'$ of $N_p$ onto itself is a local diffeomorphism, and it is called a <u>local geodesic symmetry with respect to p</u>.

Now, the manifold $(M,g)$ is said to be <u>Riemannian locally symmetric</u> if, for each $p \in M$, there is an open neighbourhood $N_p$ such that the local geodesic symmetry in $N_p$ is an isometry. It is well-known that $(M,g)$ is Riemannian locally symmetric if and only if $\nabla R = 0$.

Further, a connected manifold $(M,g)$ is said to be <u>Riemannian (globally) symmetric</u> if each point $p \in M$ is an isolated fixed point of an involutive isometry $s_p$ of $(M,g)$. (Here $s_p$ is called a <u>symmetry at p</u>.) For the tangent maps $S_p = (s_p)_{*p}$ we get easily $S_p = -\mathrm{Id}$. Hence it is easy to see that a connected manifold $(M,g)$ is Riemannian symmetric if and only if it is Riemannian locally symmetric and, for each point $p$, a local geodesic symmetry with respect to $p$ can be extended to an isometry $s_p$ of $(M,g)$. The last condition is always satisfied when $(M,g)$ is connected, simply connected and complete.

On a Riemannian symmetric space $(M,g)$, the set $\{s_p : p \in M\}$ of symmetries is uniquely determined, and for every two points $x,y \in M$ we have

$$s_x \circ s_y = s_z \circ s_x \,, \quad \text{where} \quad z = s_x(y) \qquad (1).$$

In fact, let us compare the tangent maps at the point $y$. For $v \in T_y(M)$ we get $(s_x \circ s_y)_{*y}(v) = (s_{x*})_y(-v) = -(s_{x*})_y(v)$, $(s_z \circ s_x)_{*y}(v) = (s_{z*})_z((s_{x*})_y(v)) = -(s_{x*})_y(v)$, because $(s_{y*})_y$ and $(s_{z*})_z$ are equal to $-\mathrm{Id}$ in the corresponding tangent spaces. According to Appendix B1, the both sides of (1) are equal on $M$.

Finally, let us recall that a Riemannian symmetric space is always complete and the full group of isometries on it is transitive. Moreover, if $G$ is the identity component of the full isometry group and $G_o$ is the isotropy group of $G$ at a fixed point, then $G/G_o$ is canonically a reductive homogeneous space.

## Riemannian s-structures.

We pass over to some meaningful generalizations of the previous situation.

Let $(M,g)$ be a connected Riemannian manifold. An isometry of $(M,g)$ with an isolated fixed point $x \in M$ will be called a symmetry of $(M,g)$ at $x$. Clearly, if $s_x$ is a symmetry of $(M,g)$ at $x$, then the tangent map $S_x = (s_x)_{*x}$ is an orthogonal transformation of $M_x$ having no fixed vectors (with the exception of $0$).

Here we do not require that a symmetry $s_x$ at $x$ is involutive and obviously, such a transformation is not uniquely determined, in general. Our purpose is to extend the class of Riemannian symmetric spaces to a possibly broader class of Riemannian manifolds which need not admit usual geodesic symmetries but still can admit generalized symmetries in the above sense. The following theorem shows that we have homogeneity in the most general situation:

Theorem 0.1 (F.Brickel). Let $(M,g)$ be a connected Riemannian manifold admitting at least one symmetry $s_x$ at each point $x \in M$. Then the group $I(M)$ of all isometries is transitive on $M$.

Corollary 0.2. a) $(M,g)$ admits a subordinate analytic structure
b) $(M,g)$ is complete.

Theorem 0.1 will be a consequence of the following more specific result:

Theorem 0.3. Let $(M,g)$ be a connected Riemannian manifold and $I(M)$ its full isometry group provided with the compact-open topology. Let $\{s_x : x \in M\}$ be a family of symmetries on $(M,g)$ (one symmetry at each point). Then the closure $\mathrm{Cl}(\{s_x\})$ in $I(M)$ of the group generated by the set $\{s_x : x \in M\}$ is a transitive Lie group of transformations.

<u>Proof.</u> (Cf. [LO]). It is well-known that the full isometry group $I(M)$ of a connected Riemannian manifold $(M,g)$ is a Lie group with respect to the compact-open topology; in particular, it is locally compact. For the proof of the last fact, the following intermediate result is often used (see [KN I] for instance): let $\{g_n\}$ be a sequence in $I(M)$ such that $g_n(x)$ is convergent for some $x \in M$, then a subsequence $\{g_{n_i}\}$ of $\{g_n\}$ exists which is convergent in $I(M)$.

The same result is true if we replace $I(M)$ by a closed subgroup $G \subset I(M)$. Now, let $G(x)$ be the G-orbit of $x$ and $y$ be a point of the closure of $G(x)$. We conclude that $y = g(x)$ for some $g \in G$. Hence we obtain

<u>Lemma.</u> If $G \subset I(M)$ is a closed subgroup then all G-orbits are closed in $M$.

In the following, we shall put $G = \mathrm{Cl}(\{s_x\})$. Let $x \in M$ be a fixed point, and $U$ a normal neighbourhood of $x$ with radius $a$. Let $y$ be any point in $U$ and let $r$ be the distance from $x$ to the G-orbit $G(y)$ of $y$; $r = \operatorname{Inf}_{f \in G} d(x,f(y))$.

Because the orbit $G(y)$ is closed, it is easy to show that for some $z \in G(y)$ we have $r = d(x,z)$. Suppose $r$ to be positive, then there is a unique geodesic segment joining $x$ and $z$ with length $r$. Let $w$ be a point of this geodesic segment between $x$ and $z$, and consider the effect of the symmetry $s_w$ at $w$ on $z$. Clearly $s_w(z)$ belongs to $G(y)$ because $z \in G(y)$, and $s_w(z) \neq z$.

Since the points $x$, $z$, $w$ and $s_w(z)$ are all in $U$, and the triangle inequality holds for any geodesic triangle in $U$, we have $d(x,s_w(z)) < d(x,w) + d(w,s_w(z)) = d(x,w) + d(w,z) = d(x,z) = r$, a contradiction to the fact that $r = d(x,G(y))$. Thus we have $r = 0$ and hence $x \in G(y)$. Consequently, $y \in G(x)$ and since $y$ was an arbitrary point in $U$ we have $U \subset G(x)$. Thus $G(x)$ contains an open neighbourhood of $x$, and because $G$ is transitive on each orbit, $G(x)$ is open. Because $G(x)$ is also closed, we get $G(x) = M$ and $G$ is transitive on $M$.

Finally, $G$ is a Lie group because $G \subset I(M)$ is closed. □

Corollary 0.2 states the well-known properties of Riemannian homogeneous spaces. We shall give an outline of the proof.

<u>a)</u>: Consider the Lie group $G = I(M)$ and its isotropy subgroup $H$ at a point $o \in M$. Then $G$ and $M \approx G/H$ admit subordinate analy-

tic structures such that the action of $G$ on $G/H$ is analytic. Because the metric $g$ is invariant with respect to the transitive and analytic action of $G$ on $G/H$, it is analytic. (Cf. [H]).

b): Because $I(M)$ is transitive on $M$, we can choose closed balls of the constant radius $r$ to all points of $(M,g)$. Let $\{x_n\}$ be a Cauchy sequence in $M$. For a sufficiently large $N > 0$ the ball $B(x_N; r)$ contains all $x_n$ for $n > N$ and thus $\{x_n\}$ converges in $B(x_N; r)$. □

In the future, it will be convenient to make the following

Definition 0.4. A family $\{s_x : x \in M\}$ of symmetries on a connected Riemannian manifold $(M,g)$ is called a (Riemannian) s-structure on $(M,g)$.

In this terminology, Theorem 0.1 can be stated as follows: If a connected Riemannian manifold $(M,g)$ admits at least one s-structure, then it is a homogeneous Riemannian space.

Definition 0.5. An s-structure $\{s_x : x \in M\}$ is said to be of order $k$ (where $k \geq 2$ is an integer) if $(s_x)^k = \mathrm{id}$ for all $x \in M$ and $(s_x)^\ell \neq \mathrm{id}$ for $\ell < k$.

Obviously a Riemannian manifold is symmetric if and only if it admits an s-structure of order 2 (such an s-structure, if it exists, is uniquely defined). Our natural question now is whether the Riemannian manifolds admitting s-structures of finite order are more special then those admitting general s-structures. The answer is negative according to a result by A.W.Deicke:

Theorem 0.6. If a connected Riemannian manifold admits an s-structure then it also admits an s-structure of finite order.

The proof is based on the following:

Lemma 0.7. Let $s_p$ be a symmetry of $(M,g)$ at $p \in M$. Denote by $Cl(s_p)$ the closure in $I(M,p)$ of the group generated by $s_p$. (Here $I(M,p)$ denotes the isotropy subgroup of $I(M)$ at $p$.) Then a symmetry $s_p'$ of finite order exists in $Cl(s_p)$.

Proof. We can suppose $s_p$ to be of infinite order. Denote by $S_p = (s_p)_{*p}$ the tangent map of $s_p$ at $p$ and let $I(M_p)$ be the set

$\{g_{*p}: g \in I(M,p)\}$. $I(M_p)$ is a closed subgroup of the orthogonal group $O(M_p)$ and $S_p \in I(M_p)$. The essential step is to prove that the closure $Cl(S_p)$ in $I(M_p)$ of the group generated by $S_p$ contains an orthogonal transformation $S_p'$ such that

(i) $S_p'$ is of finite order

(ii) all eigenvalues of $S_p'$ are different from 1.

Since the group $Cl(S_p)$ is compact and abelian, it is of the form $T^k \times F$, where $T^k$ is a k-dimensional torus $(k \geq 1)$ and $F$ is a finite cyclic group of order say $\ell$. Let $e$ denote the unit element of $F$. Then $T'^k = T^k \times \{e\}$ is the identity component in $Cl(S_p)$. Let $\mathcal{A}$ denote the set of all elements of $Cl(S_p)$ satisfying (ii). Then $\mathcal{A} \ni S_p$ is open and there is a neighbourhood $\mathcal{V}$ of the identity in $T'^k$ such that $S_p \cdot \mathcal{V} \subset \mathcal{A}$. Obviously, the set $\mathcal{V}^{(\ell)} = \{v^\ell : v \in \mathcal{V}\}$ is a neighbourhood of the identity in $T'^k$, too. Now, $(S_p)^\ell \in T'^k$ and thus the set $(S_p)^\ell \mathcal{V}^{(\ell)}$ is a neighbourhood of $(S_p)^\ell$ in $T'^k$. It is well-known that the set of all elements of finite order in $T'^k$ is dense and thus there is $v_o \in \mathcal{V}$ such that $(S_p)^\ell (v_o)^\ell$ is an element of finite order. Consequently, $S_p v_o$ is of finite order and, at the same time, $S_p v_o \in \mathcal{A}$. Thus we can put $S_p' = S_p v_o$.

Now, because $S_p' \in Cl(S_p)$, there exists an isometry $s_p' \in Cl(s_p)$ such that $S_p' = (s_p')_{*p}$. According to (ii), $S_p'$ has the null vector as a unique fixed vector in $M_p$ and hence $s_p'$ has the point $p$ as an isolated fixed point. (We use the normal coordinates.) Thus $s_p'$ is a symmetry of finite order. □

The proof of Theorem 0.6 follows easily from Theorem 0.1: for each $x \in M$ we take an isometry $g_x \in I(M)$ such that $g_x(p) = x$. Then we can define an s-structure $\{s_x' : x \in M\}$ of finite order in such a way that $s_p'$ is a symmetry of finite order at $p$, and $s_x' = g_x \circ s_p' \circ g_x^{-1}$ for $x \in M$. □□

The previous results are the major part of what is known in the general situation. (See also Note 3.) If we want to develop deeper analogies to the theory of the Riemannian symmetric spaces, a more special structure is needed. Fortunately, we have not made use of the property (1) so far. We shall study the corresponding class of spaces in the next paragraph. We shall also see that this class actually extends that of Riemannian symmetric spaces.

## Regular s-structures.

In the following, a Riemannian manifold with an s-structure is always considered as a real analytic manifold. (Cf. 0.2.)

Definition 0.8. An s-structure $\{s_x\}$ on a Riemannian manifold $(M,g)$ is said to be regular if it satisfies the rule

$$s_x \circ s_y = s_z \circ s_x \ , \quad z = s_x(y) \qquad (1)$$

for every two points $x, y \in M$.

Given an s-structure $\{s_x : x \in M\}$ on $(M,g)$ we shall always denote by $S$ the tensor field of type $(1,1)$ defined by $S_x = (s_{x*})_x$ for all $x \in M$. We have the following characterization of the regularity:

Proposition 0.9. An s-structure $\{s_x\}$ on a connected $(M,g)$ is regular if and only if the tensor field $S$ is invariant with respect to all symmetries $s_x$, i.e.

$$s_{x*}(S) = S \ , \quad x \in M \qquad (2).$$

Proof. Passing over to the tangent maps in Formula (1), we obtain $(s_{x*})_y \circ S_y = S_z \circ (s_{x*})_y$ at the point $y$, hence $s_{x*} \circ S = S \circ s_{x*}$ holds on the tangent bundle $T(M)$, and this is the relation (2). Conversely, because an isometry is uniquely defined by a single tangent map, the relation (1) follows from (2), too. □

Examples 0.10.

1. The family of geodesic symmetries on a Riemannian symmetric space is a regular s-structure.
2. Let $M$ be the euclidean plane and $\varphi$ an angle between $0$ and $2\pi$. Let $s_x$ be the rotation around $x$ with the angle $\varphi$ for all $x \in M$. Then $\{s_x\}$ is a regular s-structure.

Let us remark that Example 2 is not obvious if we try to check the regularity directly but it becomes evident by Proposition 0.9.

Our next aim is to prove two theorems on regular s-structures:

Theorem 0.11. Every regular Riemannian s-structure is analytic. More precisely: If $(M,g)$ admits a regular s-structure $\{s_x\}$, and if $(M,g)$ is equipped with its canonical analytic structure then the map $(x,y) \longmapsto s_x(y)$ is analytic.

**Theorem 0.12.** If $(M,g)$ admits a regular s-structure then it also admits a regular s-structure of finite order.

We shall require some lemmas:

**Lemma 0.13.** Let $\{s_x\}$ be a regular s-structure on $(M,g)$. Then the tensor field $S$ is invariant by the group $Cl(\{s_x\})$, and for any point $p \in M$ the symmetry $s_p$ belongs to the centre of the isotropy subgroup $Cl(\{s_x\}, p)$.

Proof. The invariance of $S$ and also the relation $s_p \in Cl(\{s_x\},p)$ is obvious (Proposition 0.9). Now, let $g \in Cl(\{s_x\},p)$, then the property $g_*(S) = S$ implies $g_{*p} \circ S_p = S_p \circ g_{*p}$ and $(g \circ s_p \circ g^{-1})_{*p} = g_{*p} \circ S_p \circ g_{*p}^{-1} = S_p = (s_{p*})_p$. Hence $g \circ s_p \circ g^{-1} = s_p$. □

**Lemma 0.14.** Let $G$ be a transitive Lie group of isometries of $(M,g)$ and let $G(p)$ denote the isotropy group at $p \in M$. Suppose that there is a symmetry $s_p$ at $p$ which is contained in the centre of $G(p)$. Then there is exactly one s-structure $\{s_x : x \in M\}$ with the initial value $s_p$ and such that its tangent tensor field $S$ is invariant by the group $G$. The last s-structure is regular.

Proof. For each $x \in M$ define $s_x = g \circ s_p \circ g^{-1}$, where $g \in G$ is any element such that $g(p) = x$. Then $s_x$ is a symmetry at $x$ which is independent of the choice of $g$ and thus the family $\{s_x : x \in M\}$ is well-defined. Further, for any $y \in M$ and $g \in G$ we have (using an auxiliary element $g' \in G$ such that $g'(p) = y$):

$$g \circ s_y \circ g^{-1} = g \circ g' \circ s_p \circ (g')^{-1} \circ g^{-1} = (g \circ g') \circ s_p \circ (g \circ g')^{-1} = s_{g(y)} \qquad (3).$$

Looking at the tangent maps we get $g_{*y} \circ S_y = S_{g(y)} \circ g_{*y}$, i.e., the tensor field $S$ is invariant by $G$. On the other hand, the $G$-invariant tensor field $S$ is uniquely determined by its initial value $S_p$ and thus the wanted s-structure $\{s_x\}$ is uniquely determined. Since $s_x \in G$ for each $x \in M$, the s-structure $\{s_x\}$ is regular by (3). □

Proof of Theorem 0.11: Let $\{s_x\}$ be a regular s-structure on $(M,g)$, and $p \in M$ a fixed point. According to 0.3 the group $G = Cl(\{s_x\})$ is transitive on $M$ and according to the proof of 0.2, there are analytic structures on $G$ and on $G/G(p) \approx M$ (arising from the compact-

-open topology) such that the action of $G$ on $M$ is analytic. Now, there is a neighbourhood $U$ of $p$ and an analytic section $h: U \to G$ of the fibre bundle $\pi: G \to M$. According to the previous lemmas, we have for each $x \in U$: $s_x = h(x) \circ s_p \circ h^{-1}(x)$. Now, $s_p$ is analytic because $s_p \in G$, and hence the map $(x,y) \mapsto s_x(y)$ is analytic on $U \times M$. Finally, we conclude that this map is analytic on $M \times M$. □□

From 0.11 we get

**Corollary 0.15.** For a regular s-structure $\{s_x\}$ the tangent tensor field $S$ is always analytic.

Proof of Theorem 0.12. Suppose that $(M,g)$ admits a regular s-structure $\{s_x\}$ and let $s'_p \in Cl(s_p)$ be a symmetry of finite order. Then $s'_p$ belongs to the centre of $Cl(\{s_x\},p)$ because $s_p$ does. According to Theorem 0.3 and Lemma 0.14 there is a regular s-structure $\{s'_x\}$ on $M$ with the initial value $s'_p$. Because each $s'_x$ is of the form $g \circ s'_p \circ g^{-1}$, all symmetries $s'_x$, $x \in M$, are of the same order as $s'_p$. □□

Now we shall introduce some of the basic concepts for our theory.

**Definition 0.16.** A generalized symmetric Riemannian space is a connected Riemannian manifold $(M,g)$ admitting a regular s-structure.

**Definition 0.17.** A Riemannian manifold $(M,g)$ is said to be k-symmetric ($k \geq 2$) if it admits a regular s-structure of order $k$.

According to Theorem 0.12 every g.s. Riemannian space is k-symmetric for some $k \geq 2$. Hence we can introduce

**Definition 0.18.** Order of a generalized symmetric Riemannian space $(M,g)$ is the least integer $k \geq 2$ for which $(M,g)$ is k-symmetric.

According to 0.10, the euclidean plane is trivially k-symmetric for each $k \geq 2$, thus its order is 2. On the other hand, g.s. Riemannian spaces of order 2 are nothing but the Riemannian symmetric spaces. So, to justify our general theory, we need to prove the existence of spaces of order $k > 2$. This is the purpose of the next paragraph.

Let us remark that the only g.s. Riemannian spaces of dimension 2 are those of constant curvature because of homogeneity. Thus the first non-trivial example can occur in the dimension $n = 3$.

## E x i s t e n c e   t h e o r e m s .

We shall start with some preparations. Let $(M,g)$ be a Riemannian homogeneous space, i.e., a homogeneous manifold $M = G/H$ equipped with a G-invariant Riemannian metric. Let $\pi : G \longrightarrow M$ denote the canonical projection, $e$ the unit element of $G$ and $o = \pi(H)$ the origin of $M$.

Proposition 0.19. Let $G$ admit an automorphism $\sigma$ such that

(i) $(G^\sigma)^\bullet \subset H \subset G^\sigma$, where $G^\sigma$ is the fixed point set of $\sigma$ and $(G^\sigma)^\bullet$ is its identity component,

(ii) $\sigma^k$ = identity, and $\sigma^\ell \neq$ identity for $\ell < k$,

(iii) the transformation $s$ of $M$ determined by $\pi \circ \sigma = s \circ \pi$ is metric preserving at the origin $o$ of $M$.

Then $M$ is a k-symmetric Riemannian space.

Proof. Suppose first that $G$ acts effectively on the coset space $G/H$. We shall identify the elements of $G$ with the corresponding transformations of $M = G/H$. Choose $g \in G$ and $x \in M$, then $x = \pi(g')$ for some $g' \in G$. Now, $(s \circ g \circ s^{-1})(x) = (s \circ g \circ s^{-1} \circ \pi)(g') = (s \circ g \circ \pi)(\sigma^{-1}(g')) = (s \circ \pi)(g\sigma^{-1}(g')) = (\pi \circ \sigma)(g\sigma^{-1}(g')) = \pi(\sigma(g)g') = \sigma(g)[\pi(g')] = \sigma(g)(x)$. Hence we get

$$s \circ g \circ s^{-1} = \sigma(g) \quad \text{for each } g \in G \qquad (4).$$

Thus the map $s \circ g \circ s^{-1}$ is always an isometry. Choose $x \in M$ and $g \in \pi^{-1}(x)$. Then the tangent maps $s_{*o}^{-1}: M_o \longrightarrow M_o$, $g_{*o}: M_o \longrightarrow M_x$ and $s_{*x} \circ g_{*o} \circ s_{*o}^{-1}: M_o \longrightarrow M_{s(x)}$ are all metric preserving. Consequently, the map $s_{*x}: M_x \longrightarrow M_{s(x)}$ is metric preserving and $s$ is an isometry of $M$.

For $x \in M$ define an isometry $s_x$ of $M$ by the formula $s_x = g \circ s \circ g^{-1}$, where $g \in \pi^{-1}(x)$. Then $s_x$ is independent of the choice of $g$. In fact, for each $h \in H$ we have $s \circ h \circ s^{-1} = \sigma(h) = h$, and hence $h \circ s \circ h^{-1} = s$.

It is obvious that $(s_x)^k$ = identity for each $x \in M$, and that $k$ is the minimum number with this property. We have to prove that $x$ is an isolated fixed point of $s_x$. For, it is sufficient to prove that $o$ is an isolated fixed point of $s$. Let $X \in M_o$ be such that $s_{*o}(X) = X$ and $\tilde{X} \in G_e$ such that $\pi_{*e}(\tilde{X}) = X$. Then $\sigma_{*e}(\tilde{X}) = \tilde{X} + \tilde{Z}$, where $\tilde{Z} \in H_e$. Now $\sigma_{*e}(\tilde{Z}) = \tilde{Z}$ and $(\sigma_{*e})^k(\tilde{X}) = \tilde{X} + k\tilde{Z} = \tilde{X}$, because $(\sigma_{*e})^k$ = identity. Thus $\tilde{Z} = 0$ and $\tilde{X}$ is a fixed point of $\sigma_{*e}$. We deduce $\tilde{X} \in H_e$ and $X = 0$. Because $s_{*o}$ has no non-zero fixed vectors and $s$ is an isometry of $M$, we conclude that $o$ is an isolated fixed point of $s$.

To prove the formula $s_x \circ s_y = s_{s_x(y)} \circ s_x$, recall once again that $s \circ g \circ s^{-1} = \sigma(g)$. Put $s_x = g \circ s \circ g^{-1}$, $s_y = g' \circ s \circ (g')^{-1}$, where $x = g(o)$ and $y = g'(o)$. Then $(g \circ s \circ g^{-1} \circ g' \circ s^{-1})(o) = s_x(g'(o)) = s_x(y)$; on the other hand, $g \circ s \circ g^{-1} \circ g' \circ s^{-1} = g \circ \sigma(g^{-1} g') = g''$ belongs to $G$. Consequently, $s_x \circ s_y = g \circ s \circ g^{-1} \circ g' \circ s \circ (g')^{-1} = g'' \circ s \circ (g'')^{-1} \circ g \circ s \circ g^{-1} = s_{s_x(y)} \circ s_x$.

Now, if $G$ does not act effectively on $G/H$, the elements $g \in G$ acting as the identity transformation on $G/H$ form a closed normal subgroup $N$ of $G$ such that $N \subset H$. Here $\sigma$ induces an automorphism $\tilde{\sigma}$ of the Lie group $G/N$. We can write $M = G/N \big/ H/N$, and the conditions of Proposition 0.19 are satisfied for $G/N$, $H/N$ and $\tilde{\sigma}$. Moreover, the group $G/N$ acts effectively on $M$. Hence the result follows. □

-------

Let $M = G/H$ be a homogeneous space, and denote by $\underline{g}, \underline{h}$ the Lie algebras of $G$ and $H$ respectively. We shall identify $\underline{g}$ with $G_e$ and $\underline{h}$ with $H_e$.

Suppose that there is a subspace $\underline{m} \subset \underline{g}$ such that $\underline{g} = \underline{h} + \underline{m}$ (direct sum) and $\mathrm{ad}(h)\underline{m} \subset \underline{m}$ for all $h \in H$. (The homogeneous space $G/H$ is then said to be reductive, see Chapter I.) The projection $\pi : G \longrightarrow G/H$ induces the tangent map $\pi_{*e} : \underline{g} \longrightarrow M_o$ which is trivial on $\underline{h}$ and a linear isomorphism on $\underline{m}$. Having a fixed $\underline{m}$, we can identify $\underline{m}$ with $M_o$ via $\pi_*$.

Now, suppose that there exists a bi-invariant Riemannian metric $\mathcal{B}$ on $G$, or equivalently, a positive scalar product $B$ on the Lie algebra $\underline{g}$ which is invariant with respect to the group $\mathrm{Int}(\underline{g}) = \mathrm{ad}(G)$ of all inner automorphisms of $\underline{g}$. Then we have a canonical reductive decomposition $\underline{g} = \underline{h} + \underline{m}$, where $\underline{m}$ is the orthogonal complement of $\underline{h}$ with respect to $B$. In fact, $B$ and $\underline{h}$ are both $\mathrm{ad}(H)$-invariant and hence $\underline{m}$ is $\mathrm{ad}(H)$-invariant, too. The restriction of $B$ to $\underline{m}$ is a positive scalar product. Hence we obtain a positive scalar product $g_o$ in $M_o$, which is invariant with respect to the linear isotropy representation of $H$ in the tangent space $M_o = (G/H)_o$. Thus $g_o$ defines a unique $G$-invariant Riemannian metric $g$ on $M$. We shall say that $g$ is <u>induced</u> by the bi-invariant metric $\mathcal{B}$ or by the ad(G)-invariant scalar product $B$.

If the group $G$ is compact and semi-simple, we always have an ad(G)-invariant positive scalar product on $\underline{g}$ - namely the negative of the Killing form, $B(X,Y) = -\mathrm{tr}(\mathrm{ad}X \circ \mathrm{ad}Y)$.

-------

We shall recall a standard construction of certain Riemannian symmetric spaces. Let $G$ be a compact connected Lie group and let $G^*$ be the diagonal of the direct product $G \times G$. The coset space $(G \times G)/G^*$ is diffeomorphic to $G$ where the diffeomorphism is determined by the map $(x_1,x_2) \mapsto x_1x_2^{-1}$ in one direction, and by the map $y \mapsto (y,e)$ in the other direction. The corresponding action of $G \times G$ on $G$ is given by $(x_1,x_2)y = x_1yx_2^{-1}$, the isotropy group at the origin is again $G^*$. Now, let $\sigma : G \times G \longrightarrow G \times G$ be the involutive automorphism given by $\sigma(x_1,x_2) = (x_2,x_1)$. The fixed points of $\sigma$ are just the elements of $G^*$ and thus $\sigma$ induces a transformation $s: G \longrightarrow G$. Obviously, $s(y) = y^{-1}$ for $y \in G$.

Now, take a bi-invariant Riemannian metric $\Phi$ on $G$; such a metric always exists. Then $\Phi$ is invariant with respect to the action of $G \times G$ on $G$. Also, $\Phi$ is preserved by the map $s$ at the identity because $s_{*e}(X) = -X$ holds for each $X \in G_e$. According to Proposition 0.19, $(G, \Phi)$ is a Riemannian symmetric space.

-------

We shall now consider the homogeneous space $G^{k+1}/\Delta G^{k+1}$, where $G^{k+1}$ is the direct product of $k+1$ copies of $G$ and $\Delta G^{k+1}$ is the diagonal of $G^{k+1}$. $G^{k+1}/\Delta G^{k+1}$ is diffeomorphic to $G^k$ via the map

$$\pi : (x_1,\dots,x_{k+1}) \longmapsto (x_1x_{k+1}^{-1},\dots,x_kx_{k+1}^{-1}) \qquad (5)$$

and the corresponding action of $G^{k+1}$ on $G^k$ is given by

$$(x_1,\dots,x_{k+1})(y_1,\dots,y_k) = (x_1y_1x_{k+1}^{-1},\dots,x_ky_kx_{k+1}^{-1}).$$

Each tangent vector at the origin of $G^k$ has a unique expression $(X_1,\dots,X_k)$, where $X_1,\dots,X_k \in G_e$.

Consider the map $\sigma : G^{k+1} \longrightarrow G^{k+1}$ defined by

$$\sigma(x_1,\dots,x_{k+1}) = (x_{k+1},x_1,\dots,x_k) \qquad (6).$$

Then $\sigma$ is an automorphism such that $\sigma^{k+1} = \mathrm{id}$, and $(G^{k+1})^{\sigma} = \Delta G^{k+1}$. Thus the conditions (i),(ii) of Proposition 0.19 are fulfilled (if we replace k, G, H, M by k+1, $G^{k+1}$, $\Delta G^{k+1}$ and $G^k$ respectively). The transformation $s\colon G^k \longrightarrow G^k$ defined by $\pi \circ \sigma = s \circ \pi$ can be described explicitly by

$$s\colon (y_1,\dots,y_k) \longmapsto (y_k^{-1}, y_1 y_k^{-1},\dots,y_{k-1}y_k^{-1}) \qquad (7).$$

Here the origin $o \in G^k$ is an isolated fixed point of s.

Let $\Phi$ be a bi-invariant Riemannian metric on G (such a metric exists because G is compact), then the direct sum $\Phi^{k+1} = \underbrace{\Phi + \dots + \Phi}_{(k+1)\text{-times}}$ is a bi-invariant Riemannian metric on $G^{k+1}$. It induces a $G^{k+1}$-invariant Riemannian metric on $G^k$; the latter will be denoted by $\Phi^{[k]}$.

Consider the scalar product $\Phi_e^{k+1}$ at the identity of $G^{k+1}$. It is invariant with respect to the automorphism $\sigma_{*e}\colon (X_1,\dots,X_{k+1}) \mapsto (X_{k+1},X_1,\dots,X_k)$ of the Lie algebra $\underline{g}^{k+1}$ (see Formula (6)), and the subalgebra $\Delta\underline{g}^{k+1} \subset \underline{g}^{k+1}$ is also $\sigma_*$-invariant. Hence it follows that the subspace $\underline{m}$ in the orthogonal decomposition $\underline{g}^{k+1} = \Delta\underline{g}^{k+1} + \underline{m}$ is $\sigma_*$-invariant. If we identify $\underline{m}$ with the tangent space $(G^k)_o$ via $\pi_*$, we see that the tangent map $s_{*o}\colon (G^k)_o \longrightarrow (G^k)_o$ preserves the metric $\Phi^{[k]}$ at the origin $o \in G^k$. According to Proposition 0.19, $(G^k, \Phi^{[k]})$ is a (k+1)-symmetric Riemannian space; a symmetry of order k+1 at the origin is given by (7).

We shall now find an explicit formula for $\Phi^{[k]}$ at the origin $o \in G^k$. Firstly, the Lie subalgebra $\Delta\underline{g}^{k+1}$ consists of vectors of the form $(X,\dots,X)$, $X \in \underline{g}$, and its orthogonal complement $\underline{m}$ is formed by all (k+1)-tuples $(X_1,\dots,X_{k+1}) \in \underline{g}^{k+1}$ such that $X_1+\dots+X_{k+1} = 0$.

Let us identify the tangent space $(G^k)_o$ with the Lie algebra $\underline{g}^k$. Then the projection (5) has the tangent map $\pi_{*e}\colon \underline{g}^{k+1} \longrightarrow \underline{g}^k$ given by

$$\pi_{*e}(X_1,\dots,X_{k+1}) = (X_1 - X_{k+1},\dots,X_k - X_{k+1}) \qquad (8).$$

Let $\wp$ denote the (unique) linear isomorphism of $\underline{g}^k$ onto $\underline{m}$ which is compatible with $\pi_{*e}$. We see easily that, for every $X_1,\dots,X_k \in \underline{g}$, we have

$$\wp(X_1,\dots,X_k) = (X_1 - \tfrac{1}{k+1}Y,\dots, X_k - \tfrac{1}{k+1}Y, -\tfrac{1}{k+1}Y) \qquad (9)$$

where $Y = X_1 + \dots + X_k$.

Now, for the induced metric $\Phi^{[k]}$ we obtain

$$\Phi_o^{[k]}((X_1,\dots,X_k),(X_1,\dots,X_k)) = \Phi_e^{k+1}((X_1 - \tfrac{1}{k+1}Y,\dots,X_k - \tfrac{1}{k+1}Y, -\tfrac{1}{k+1}Y),$$

$$(X_1 - \tfrac{1}{k+1}Y,\dots,X_k - \tfrac{1}{k+1}Y, -\tfrac{1}{k+1}Y)) = \sum_{i=1}^{k}\Phi(X_i - \tfrac{1}{k+1}Y, X_i - \tfrac{1}{k+1}Y) +$$

$$+ \frac{1}{(k+1)^2}\Phi(Y,Y) = \sum_{i=1}^{k}\Phi(X_i,X_i) - \tfrac{1}{k+1}\Phi(X_1 +\dots+ X_k, X_1 +\dots+ X_k) =$$

$$= \tfrac{1}{k+1}(k\sum_{i=1}^{k}\Phi(X_i,X_i) - 2\sum_{i<j}\Phi(X_i,X_j)) =$$

$$= \tfrac{1}{k+1}\Big[\sum_{i=1}^{k}\Phi(X_i,X_i) + \sum_{1\le i<j\le k}\Phi(X_i - X_j, X_i - X_j)\Big] .$$

This metric coincides, up to a constant factor, with that studied by Ledger and Obata in [LO]. Following [LO], we shall now prove that $(G^k, \Phi^{[k]})$ is not locally symmetric, in general.

If $X$ is a vector field on $G$, then $X^{(i)}$ will denote the vector field on $G^k$ such that, for each $\tilde{x}\in G^k$, $\tilde{x} = (x_1,\dots,x_k)$, we have $X^{(i)}_{\tilde{x}} = (0_{x_1},\dots, 0_{x_{i-1}}, X_{x_i}, 0_{x_{i+1}},\dots, 0_{x_k})$.

Let $\{X_\alpha\}$, $\alpha = 1,\dots, r$, be an orthonormal basis of left-invariant vector fields on $G$ with respect to $\Phi$. Then $\{X^{(i)}_\alpha\}$, $\alpha = 1,\dots,r$, $i = 1,\dots,k$, is a basis of the module $\mathcal{X}(G^k)$ over $\mathcal{F}(G^k)$, and the vector fields $X^{(i)}_\alpha$ are invariant with respect to the action of the subgroup $G^{k*} \subset G^{k+1}$ formed by all the elements $(x_1,\dots,x_k,e)$. Hence all of the functions $\Phi^{[k]}(X^{(i)}_\alpha, X^{(j)}_\beta)$ are constants. More specifically, for left-invariant vector fields $X$, $Y$ on $G$ we have

$$\Phi^{[k]}(X^{(i)},Y^{(j)}) = -\frac{1}{k+1}\,\Phi(X_e,Y_e) \quad \text{for } i\neq j,\ i,j = 1,\dots,k$$

$$\Phi^{[k]}(X^{(i)},Y^{(i)}) = (k/k+1)\,\Phi(X_e,Y_e), \quad i = 1,\dots,k. \qquad (10)$$

Let $\nabla$ denote the Riemannian connection and $R$ the curvature tensor field associated with $\Phi^{[k]}$. It follows from the previous remarks that the connection $\nabla$ is completely determined by the formula

$$\Phi^{[k]}(\nabla_{X^{(i)}_\alpha}X^{(j)}_\beta, X^{(\ell)}_\gamma) = \tfrac{1}{2}\{\Phi^{[k]}([X^{(i)}_\alpha, X^{(j)}_\beta], X^{(\ell)}_\gamma) +$$

$$+ \Phi^{[k]}([X^{(\ell)}_\gamma, X^{(i)}_\alpha], X^{(j)}_\beta) + \Phi^{[k]}([X^{(\ell)}_\gamma, X^{(j)}_\beta], X^{(i)}_\alpha)\} \qquad (11).$$

If $X$ and $Y$ are left-invariant vector fields on $G$, then we can check easily

$$\nabla_{X^{(i)}} Y^{(j)} = \frac{1}{2(k+1)}([X,Y]^{(j)} - [X,Y]^{(i)}) \quad \text{for} \quad i \neq j$$
$$\nabla_{X^{(i)}} Y^{(i)} = \tfrac{1}{2}[X,Y]^{(i)} \quad . \tag{12}$$

A straightforward calculation then gives, for $i \neq j$,

$$(\nabla_{X^{(i)}} R)(X^{(i)}, X^{(j)})Y^{(j)} = \frac{1}{8(k+1)^3}[(2-k^2)((\mathrm{ad}X)^3Y)^{(i)} + k((\mathrm{ad}X)^3Y)^{(j)}].$$

Thus, for $k > 1$, $r > 1$, the relation $\nabla R = 0$ implies that the Lie algebra of $G$ is nilpotent and hence abelian, since $G$ is compact. Supposing that $G$ is non-abelian, we obtain $\nabla R \neq 0$. We summarize:

Theorem 0.20. If $G$ is a connected, compact and non-abelian Lie group, $\Phi$ a bi-invariant metric on $G$ and $k > 2$ an integer, then the Riemannian homogeneous space $(G^k/\Delta G^k, \Phi^{[k-1]})$ is $k$-symmetric but not symmetric.

-------

We are now going to prove a stronger result, namely, the existence of spaces of order $k$. We shall make use of the previous class of Riemannian spaces.

Let $G/H$ be a Riemannian homogeneous space with the metric $g$ and let $H$ be compact. Let $\tau : G \longrightarrow I(G/H,g)$ be the representation of $G$ by the isometries. Then it is a standard fact that $\tau(G)$ is a closed Lie subgroup of the isometry group $I(G/H,g)$, and hence the Lie algebra $\tau(\underline{g})$ of $\tau(G)$ is a subalgebra of $\underline{i}(G/H,g)$. (See e.g. [H].) In the particular case of a homogeneous manifold $(G^{k+1}/\Delta G^{k+1}, \Phi^{[k]})$, the Lie algebra $\tau(\underline{g}^{k+1})$ is isomorphic with the factor algebra $\underline{g}^{k+1}/\Delta\underline{z}^{k+1}$, where $\underline{z}$ is the center of $\underline{g}$.

Proposition 0.21. Consider a homogeneous Riemannian manifold $(G^{k+1}/\Delta G^{k+1}, \Phi^{[k]}) = (G^k, \Phi^{[k]})$, where a) $G$ is compact and simple, b) $\Phi$ is determined by a negative multiple of the Killing form of $\underline{g}$. Suppose that $\tau(G^{k+1})$ is the identity component of the full isometry group $I(G^k, \Phi^{[k]})$. Then $(G^k, \Phi^{[k]})$ is not $\ell$-symmetric for any $\ell < k + 1$.

Proof. According to the previous remarks, $\underline{g}^{k+1}$ is isomorphic to the Lie algebra $\underline{i}(G^k, \Phi^{[k]})$ and $G^{k+1}$ is locally isomorphic to $I(G^k, \Phi^{[k]})$.

Let $r$ be an isometry of $(G^k, \Phi^{[k]})$ with the isolated fixed point $o = (e,\ldots,e)$ and such that $r^\ell$ = identity. Define an automorphism $\wp$ of the group $I(G^k, \Phi^{[k]})$ by the formula $\wp(g) = r\circ g\circ r^{-1}$. Then $\wp$ induces an automorphism $d\wp$ of the Lie algebra $\underline{g}^{k+1}$. Identifying $\underline{g}^{k+1}$ with $(G^{k+1})_e$, we get

$$\pi_{*e}\circ d\wp = r_{*o}\circ\pi_{*e} \quad \text{on} \quad (G^{k+1})_e \qquad (13).$$

Hence and from (8) we obtain

$$d\wp(\Delta\underline{g}^{k+1}) \subset \Delta\underline{g}^{k+1} \qquad (14).$$

$\underline{g}^{k+1}$ is a direct sum of simple subalgebras $\underline{g}^{*(i)}$, $i = 1,\dots,k+1$, all of them being canonically isomorphic to $\underline{g}$. The automorphism $d\wp$ induces a permutation $\nu$ of the indices $1,\dots,k+1$ such that $d\wp(\underline{g}^{*\nu(i)}) = \underline{g}^{*(i)}$, $i = 1,\dots,k+1$. Denote by $\varphi_i$ the restriction of $d\wp$ to $\underline{g}^{*\nu(i)}$. Then we can write $d\wp(X_1,\dots,X_{k+1}) = (\varphi_1(X_{\nu(1)}),\dots,\varphi_{k+1}(X_{\nu(k+1)}))$, where $X_i\in\underline{g}$ for $i = 1,\dots,k+1$. In particular, $d\wp(X,\dots,X) = (\varphi_1(X),\dots,\varphi_{k+1}(X))$. From (14) we have $\varphi_1 = \varphi_2 = \dots = \varphi_{k+1}$ under the canonical identification $\underline{g}^{*(1)} = \dots = \underline{g}^{*(k+1)} = \underline{g}$. We obtain a unique automorphism $\varphi\colon \underline{g}\longrightarrow\underline{g}$ such that

$$d\wp(X_1,\dots,X_{k+1}) = (\varphi(X_{\nu(1)}),\dots,\varphi(X_{\nu(k+1)})) \qquad (15).$$

Now, we shall make use of a result by J.Winter, [Wi].

<u>Lemma 0.22.</u> An automorphism A of a non-solvable Lie algebra $\underline{g}$ leaves fixed an element X such that adX is not nilpotent.

Let $X \neq 0$ be a fixed vector of $\varphi\colon \underline{g}\longrightarrow\underline{g}$ and suppose $\ell < k+1$. Then the permutation $\nu$ contains a cycle $(i_1,\dots,i_m)$ of the length $m < k+1$. Consider the vector $Z = (X_1,\dots,X_{k+1})\in\underline{g}^{k+1}$ such that $X_i = X$ for $i = i_1,\dots,i_m$ and $X_i = -X$ otherwise. Clearly, $d\wp(Z) = Z$, and from (13) we get that $\pi_{*e}(Z)$ is a fixed vector of $r_{*o}$. Further, Z does not belong to the subalgebra $\Delta\underline{g}^{k+1}$ and hence $\pi_{*e}(Z)\neq 0$, a contradiction. □

<u>Proposition 0.23.</u> For $G = SO(3)$ and $\Phi(X,Y) = -(1/2)\mathrm{tr}(\mathrm{ad}X\circ\mathrm{ad}Y)$ the last condition of Proposition 0.21 is satisfied for each k.

<u>Proof.</u> In the following, the elements of $\underline{g}$ (or $\underline{g}^k$) are considered as left invariant vector fields on G (or $G^k$) respectively. First of all, there is a basis $\{X_1,X_2,X_3\}$ of $\underline{g}$ such that $[X_1,X_2] = X_3$, $[X_2,X_3] = X_1$, $[X_3,X_1] = X_2$. We have $\Phi(X_\alpha,X_\beta) = \delta_{\alpha\beta}$ for $\alpha,\beta =$

$= 1,2,3$. The vectors $\{X_\alpha^{(i)}\}$, $\alpha = 1,2,3$; $i = 1,\ldots,k$, form a basis of $\underline{g}^k$. Using the formulas (12) and the multiplication in $\underline{g}$ we can get easily the following properties of the curvature tensor $R$ of $\Phi^{[k]}$:

$$R(X_\alpha^{(i)}, X_\beta^{(j)})X_\gamma^{(\ell)} = 0 \quad \text{for } \alpha \neq \beta \neq \gamma \text{ or } \alpha = \beta = \gamma,$$

$$R(X_\alpha^{(i)}, X_\beta^{(j)})X_\alpha^{(\ell)} \quad \text{and} \quad R(X_\alpha^{(i)}, X_\alpha^{(j)})X_\beta^{(\ell)} \qquad (16)$$

belong to the subspace generated by $X_\beta^{(i)}, X_\beta^{(j)}, X_\beta^{(\ell)}$.

Let $H_o$ be the identity component of the isotropy group of $I(G^k, \Phi^{[k]})$ at the origin $o$. Denote by $\underline{h}_o$ the corresponding Lie algebra. Then $\underline{h}_o$ possesses a faithful isotropy representation by the endomorphisms of $(G^k)_o$. Clearly, a necessary condition for $A \in \underline{h}_o$ is that $A(\Phi^{[k]}) = A(R) = 0$ at the origin, where $A$ acts as a derivation on the tensor algebra of $(G^k)_o$. In the following, all calculations will be done in the tangent space $(G^k)_o$. For our convenience, we shall work with the homothetic metric $\hat{\Phi}^{[k]} = (k+1)\,\Phi^{[k]}$ instead of $\Phi^{[k]}$.

Let $A \in \underline{h}_o$ and set

$$AX_\alpha^{(i)} = \sum_{\beta=1}^{3} \sum_{j=1}^{k} a\binom{i}{j}_\alpha^\beta X^{(j)}, \quad i = 1,\ldots,k; \quad \alpha = 1,2,3 \qquad (17).$$

We see that $\hat{\Phi}^{[k]}(X^{(i)}, X^{(j)}) = \begin{cases} -\delta_{\alpha\beta} & \text{for } i \neq j \\ k\,\delta_{\alpha\beta} & \text{for } i = j. \end{cases}$

The relation $(A\,\hat{\Phi}^{[k]})(X_\alpha^{(i)}, X_\beta^{(j)}) = 0$ implies that

$$k\left(a\binom{i}{i}_\alpha^\beta + a\binom{j}{j}_\beta^\alpha\right) - \sum_{\ell \neq i} a\binom{i}{\ell}_\alpha^\beta - \sum_{\ell \neq j} a\binom{j}{\ell}_\beta^\alpha = 0 \qquad (18).$$

Further, we calculate easily that

$$R(X_\alpha^{(i)}, X_\beta^{(i)})X_\alpha^{(i)} = -(1/4)X_\beta^{(i)} \quad \text{for } \alpha \neq \beta.$$

Consider the relation $(AR)(X_\alpha^{(i)}, X_\alpha^{(i)})X_\beta^{(i)} = 0$, i.e.,

$$-(1/4)AX_\beta^{(i)} = R(AX_\alpha^{(i)}, X_\beta^{(i)})X_\alpha^{(i)} + R(X_\alpha^{(i)}, AX_\beta^{(i)})X_\alpha^{(i)} +$$
$$+ R(X_\alpha^{(i)}, X_\beta^{(i)})AX_\alpha^{(i)} \qquad (19).$$

Let us substitute (17) into (19) and consider a vector $X_\gamma^{(j)}$, where $\gamma \neq \alpha, \beta$ and $j \neq i$. This vector enters into the left hand side with the coefficient $-(1/4)a\binom{i}{j}_\beta^\gamma$. According to (16), there is only one term on the right hand side the evaluation of which can involve $X_\gamma^{(j)}$, namely the term $R(X_\alpha^{(i)}, a\left\{{(i) \atop (j)}\right\}_\beta^\gamma X_\gamma^{(j)})X_\alpha^{(i)}$. Now $R(X_\alpha^{(i)}, a\left\{{(i) \atop (j)}\right\}_\beta^\gamma X_\gamma^{(j)})X_\alpha^{(i)} =$ $= a\left\{{(i) \atop (j)}\right\}_\beta^\gamma [(k+2)X_\gamma^{(i)} - X_\gamma^{(j)}]/[4(k+1)^2]$. Comparing the coefficients at $X_\gamma^{(j)}$ we get finally $a\left\{{(i) \atop (j)}\right\}_\beta^\gamma = 0$. Thus we have proved

$$a\left\{{(i) \atop (j)}\right\}_\beta^\alpha = 0 \quad \text{for} \quad i \neq j, \ \alpha \neq \beta \tag{20}$$

Substituing into (18) we get

$$a\left\{{(i) \atop (i)}\right\}_\alpha^\beta + a\left\{{(j) \atop (j)}\right\}_\beta^\alpha = 0 \quad \text{for } \alpha \neq \beta \tag{21}$$

In particular, for $i = j$ we obtain

$$a\left\{{(i) \atop (i)}\right\}_\alpha^\beta + a\left\{{(i) \atop (i)}\right\}_\beta^\alpha = 0 \tag{22}$$

and hence

$$a\left\{{(1) \atop (1)}\right\}_\alpha^\beta = a\left\{{(2) \atop (2)}\right\}_\alpha^\beta = \ldots = a\left\{{(k) \atop (k)}\right\}_\alpha^\beta \quad \text{for } \alpha \neq \beta \tag{23}$$

Now, let us compare the coefficients at $X_\beta^{(j)}$, $j \neq i$, in the relation (19). $X_\beta^{(j)}$ enters into the left hand side with the coefficient $-(1/4)\, a\left\{{(i) \atop (j)}\right\}_\beta^\beta$. As for the right hand side, $X_\beta^{(j)}$ can be involved only in the evaluation of the terms $R(a\left\{{(i) \atop (j)}\right\}_\alpha^\alpha X_\alpha^{(j)}, X_\beta^{(i)})X_\alpha^{(i)}$, $R(X_\alpha^{(i)}, a\left\{{(i) \atop (j)}\right\}_\beta^\beta X_\beta^{(j)})X_\alpha^{(i)}$, $R(X_\alpha^{(i)}, X_\beta^{(i)})(a\left\{{(i) \atop (j)}\right\}_\alpha^\alpha X_\alpha^{(j)})$. After routine calculations we obtain

$$(3k+2)a\left\{{(i) \atop (j)}\right\}_\alpha^\alpha + (k^2 - 2k)a\left\{{(i) \atop (j)}\right\}_\beta^\beta = 0, \quad i \neq j \tag{24}$$

Writing these relations for $(\alpha, \beta) = (1,2), (2,3), (3,1)$ respectively, we obtain finally

$$a\left\{{(i) \atop (j)}\right\}_\alpha^\alpha = 0 \quad \text{for} \quad i \neq j, \quad \alpha = 1,2,3 \tag{25}$$

Having $i=j$ and $\alpha = \beta$ in (18), we deduce from (25)

$$a\left\{{(i) \atop (i)}\right\}_\alpha^\alpha = 0, \quad \alpha = 1,2,3; \quad i = 1,\ldots,k \tag{26}$$

Summarizing (20),(25) and then (22),(23),(26), we see that $\underline{h}_o$ is isomorphic to $\underline{so}(3)$. Hence $H_o$ is locally isomorphic to $G = SO(3)$.

On the other hand, $H_o$ contains the image $\tau(\Delta G^{k+1})$, which is also locally isomorphic to G. Hence $H_o = \tau(\Delta G^{k+1})$, and the identity component of $I(G^k, \Phi^{[k]})$ coincides with $\tau(G^{k+1})$. □

We conclude:

Theorem 0.24. For each integer $k \geq 2$ there exists a compact generalized symmetric Riemannian space $(M,g)$ of order $k$ such that the identity component of the full isometry group is semi-simple.
In particular, for each integer $k > 2$ there is a k-symmetric Riemannian space which is not $\ell$-symmetric for $\ell = 2,\dots,k-1$.

Remark. It would be interesting to know if the last condition of Proposition 0.21 is automatically satisfied for every compact simple Lie group. M.Božek has proved by direct calculations (unpublished) that Proposition 0.23 can be extended to the case $G = SO(n)$.

## A low-dimensional example.

In the previous section we have constructed a class of g.s. Riemannian spaces of order $> 2$ which are all compact and for which the identity component of the isometry group is semi-simple. This last property of the isometry group is typical for the ordinary symmetric spaces. In fact, if $(M,g)$ is a connected and simply connected Riemannian symmetric space, then according to the de Rham decomposition theorem we have $M = M_o \times M'$, where $M_o$ is a euclidean space and $M'$ is a direct product of irreducible symmetric spaces. Now, $I^\bullet(M) = I^\bullet(M_o) \rtimes I^\bullet(M')$, and $I^\bullet(M')$ is semi-simple. (Cf. [KN II].)

This situation is no longer the unique pattern for the class of all generalized symmetric Riemannian spaces. Here there are also spaces with a different group-theoretical structure. As an illustration, we shall construct a generalized symmetric Riemannian space $(M,g)$ such that

a) $(M,g)$ is diffeomorphic to $R^3$, irreducible and of order 4,
b) the group $I^\bullet(M)$ is solvable.

Consider the Lie group $G$ of all matrices of the form

$$\begin{Vmatrix} e^{-c} & 0 & a \\ 0 & e^{c} & b \\ 0 & 0 & 1 \end{Vmatrix}$$ (the group of hyperbolic motions of the plane $R^2$).

G is solvable and diffeomorphic to $R^3$. The vector fields $X = e^{-c}(\frac{\partial}{\partial a})$, $Y = e^{c}(\frac{\partial}{\partial b})$, $Z = \frac{\partial}{\partial c}$ are left-invariant and form a basis for the Lie algebra $\underline{g}$ of G. We have $[X,Y] = 0$, $[X,Z] = X$, $[Y,Z] = -Y$. Define an invariant Riemannian metric $g$ on G by the formulas $g(X,X) = g(Y,Y) = 1$, $g(Z,Z) = \lambda^2$, $g(X,Y) = g(X,Z) = g(Y,Z) = 0$, $\lambda > 0$. Explicitly, on the underlying manifold $R^3(a,b,c)$ we have

$$ds^2 = e^{2c}da^2 + e^{-2c}db^2 + \lambda^2 dc^2 , \quad \lambda > 0.$$

The automorphism $A: \underline{g} \longrightarrow \underline{g}$ given by $AX = -Y$, $AY = X$, $AZ = -Z$ is metric-preserving and of order 4. The corresponding automorphism $\sigma$ of G is given by the formulas $a' = -b$, $b' = a$, $c' = -c$. Obviously, $\sigma$ is an isometry of the space $(R^3(a,b,c), g)$, it is of order 4, and the origin is the unique fixed point of $\sigma$. According to Proposition 0.19 (where $\pi = id$, $\sigma = s$) the space $(R^3,g)$ is 4-symmetric.

For the Riemannian connection $\nabla$ we derive easily the following formulas:

$$\begin{aligned} &\nabla_X X = -\lambda^2 Z, \quad \nabla_X Y = 0, \quad \nabla_X Z = X \\ &\nabla_Y X = 0, \quad \nabla_Y Y = \lambda^2 Z, \quad \nabla_Y Z = -Y \qquad\qquad (27) \\ &\nabla_Z X = \nabla_Z Y = \nabla_Z Z = 0 . \end{aligned}$$

Calculating the curvature tensor we get immediately that our space is irreducible (the Lie algebra generated by the curvature transformations is irreducible on each tangent space). Further, the sectional curvatures in the basic 2-directions are $K(X,Y) = \lambda^2$, $K(X,Z) = K(Y,Z) = -\lambda^2$ at each point. It is easily seen that $(X,Y)_p$, $p \in R^3$, are the only tangent 2-planes with the curvature $\lambda^2$. Hence this family must be preserved by each isometry I, and consequently, $I_* Z = \varepsilon Z$, $I_* X = X\cos\alpha + Y\sin\alpha$, $I_* Y = \varepsilon'(-X\sin\alpha + Y\cos\alpha)$, $\varepsilon, \varepsilon' = \pm 1$, where the parameter $\alpha$ is a real function on G. Also, each isometry $I \in I(R^3,g)$ is an affine transformation with respect to $\nabla$, i.e., $\nabla_{I_* U} I_* V = I_*(\nabla_U V)$ for any vector fields U,V on G. By examining the cases $U = X$, $V = Z$ and $U = Y$, $V = Z$ it follows from (27) that only 8 cases are possible, namely

$$\left.\begin{aligned} &I_* X = \delta X, \quad I_* Y = \delta' Y, \quad I_* Z = Z \\ &I_* X = \delta Y, \quad I_* Y = \delta' X, \quad I_* Z = -Z \end{aligned}\right\} \quad \delta, \delta' = \pm 1 .$$

In particular, the isometry group at the origin is finite, consisting of 8 elements. Hence we obtain:

a) The identity component of the group $I(R^3,g)$ is isomorphic to $G$ and thus solvable.
b) There are no symmetries of order 2 or 3 at the origin; therefore $(R^3,g)$ is generalized symmetric of order 4.

We shall see later (Chapter VI) that the spaces constructed above are the only simply connected g.s. Riemannian spaces of dimension 3 which are not symmetric. (Here, for the different values of $\lambda$ we obtain non-isometric spaces.) This example is also important for generalizations to higher dimensions. From this point of view, our example yields some "typical" g.s. Riemannian spaces of solvable type. (Cf. Note 1.)

## The de Rham decomposition.

The following generalizes a well-known result for Riemannian symmetric spaces:

Theorem 0.25. Let $(M,g)$ be a simply connected generalized symmetric Riemannian space and $M = M_o \times M_1 \times \ldots \times M_r$ its de Rham decomposition, where $M_o$ is a Euclidean space and $M_1,\ldots,M_r$ are irreducible. Then each factor is a generalized symmetric Riemannian space. Moreover, every regular s-structure of order $k$ on $(M,g)$ determines a regular s-structure of order $k_i$ on each $M_i$, where $k_i | k$ for $i = 0,1,\ldots,r$.

Proof. Choose a point $p \in M$ and let $M_p = V_o + V_1 + \ldots + V_r$ be the canonical orthogonal decomposition of $M_p$ into invariant subspaces with respect to the linear holonomy group $\Psi(p)$. Here $V_1,\ldots,V_r$ are irreducible with respect to $\Psi(p)$. (See [KN I], p.185.) Let $\{s_x\}$ be a regular s-structure of order $k \geq 2$ on $(M,g)$. The orthogonal transformation $S_p = (s_{p*})_p$ of $M_p$ is such that $(S_p)^k = I_p$, and the mapping $S_p - I_p$ is non-singular. Also, $S_p$ commutes with the holonomy group $\Psi(p)$ in the sense that $S_p\Psi(p)S_p^{-1} = \Psi(p)$. Thus $S_p$ leaves the canonical decomposition of $M_p$ invariant (up to an order) and $S_p(V_o) = V_o$. Further, for any $X \in M_p$ we have

$$X + S_p(X) + \ldots + S_p^{k-1}(X) = 0 \qquad (28).$$

Indeed, it holds $(S_p - I_p)(I_p + S_p + \ldots + (S_p)^{k-1}) = 0$, where $S_p - I_p$ is a non-singular transformation.

Let us consider (without the loss of generality) the subspace $V_1$. Then the subspaces $S_p(V_1)$, $S_p^2(V_1),\dots$, $S_p^{k-1}(V_1)$ are some of the components $V_i$. Let $\ell$ be the least positive integer such that $S_p^{\ell}(V_1) = V_1$ ($\ell \leq k$). The irreducible subspaces $V_1$, $S_p(V_1),\dots$ $\dots$, $S_p^{\ell-1}(V_1)$ are all different and mutually orthogonal. Now it is impossible that $k = \ell$; otherwise the relation (28) would provide an orthogonal decomposition of the null vector for any $X \in V_1$ and hence $V_1 = (0)$, a contradiction. Thus $k = m\ell$, where $m \geq 2$. We derive from (28): $X + S_p^{\ell}(X) + \dots + S_p^{\ell(m-1)}(X) = 0$ for each $X \in V_1$. Supposing $S_p^{\ell}(X) = X$ for some $X \in V_1$, $X \neq 0$, we would obtain $m.X =$ $= 0$, a contradiction. So, $S_p^{\ell}$ has in $V_1$ no fixed vector.

Let $M_1'$ be the totally geodesic submanifold of $M$ which is tangent to $V_1$ at $p$. ($M_1'$ is isometric to the factor $M_1$ of the de Rham decomposition.) Then the isometry $s_p^{\ell}$ preserves the manifold $M_1'$ and $s_p^{\ell}|_{M_1'}$ is a symmetry of $M_1'$ at $p$.

Now, $M$ is connected and $G = Cl(\{s_x\})$ is a transitive group of isometries of $M$, hence the identity component $G^{\bullet}$ is transitive on $M$, too. Let us recall that the leaves of the de Rham decomposition are maximal integral manifolds of the distributions $\mathcal{V}_o$, $\mathcal{V}_1,\dots$ $\dots$, $\mathcal{V}_r$ on $M$ obtained from $V_o$, $V_1,\dots$, $V_r$ by parallel displacements. Obviously, the connected isometry group $G^{\bullet}$ preserves all the distributions $\mathcal{V}_i$, $i = 0,1,\dots,r$. Hence, if $x \in M_1'$ and $g \in G^{\bullet}$ is such that $g(p) = x$, then $g$ leaves $M_1'$ invariant. Consequently, the maximal closed subgroup $G_1 \subset G$ leaving $M_1'$ invariant is transitive on $M_1'$. We also have $s_p^{\ell} \in G_1$.

For every $x \in M_1'$ we can write $s_x = g \circ s_p \circ g^{-1}$, where $g \in G_1$ is an element such that $g(p) = x$ (see proof of Lemma 0.14). Hence $s_x^{\ell} = g \circ s_p^{\ell} \circ g^{-1} \in G_1$ and $s_x^{\ell}$ induces a symmetry of $M_1'$ at $x$. The family $\{s_x^{\ell}|_{M_1'} : x \in M_1'\}$ is a regular s-structure of order $m$ on $M_1'$, which completes the proof. □

<u>Remark 0.26.</u> For ordinary Riemannian symmetric spaces we have a more general result: Let $M_1$ and $M_2$ be Riemannian manifolds. If their Riemannian direct product $M_1 \times M_2$ is Riemannian symmetric, then both $M_1$ and $M_2$ are Riemannian symmetric.

We do not know whether this is true for the generalized symmetric spaces, too (i.e., without the restriction to simply connected $M_i$).

## Parallel and non-parallel s-structures.

Definition 0.27. An s-structure $\{s_x\}$ on a Riemannian manifold $(M,g)$ is said to be parallel if the tensor field $S$ is parallel with respect to the Riemannian connection: $\nabla S = 0$.

Proposition 0.28. Each parallel Riemannian s-structure is regular.

Proof. Suppose $\{s_x\}$ to be a parallel s-structure on $(M,g)$. Let $p \in M$ be a fixed point and put $S' = s_{p*}(S)$. Because $\nabla S = 0$ and $s_p$ is connection-preserving, we have $\nabla S' = 0$. Now $S'_p = (s_{p*})_p(S_p) = S_p$; from the uniqueness of a parallel extension we have $S' = S$. Thus for all points $p \in M$ we get $(s_{p*})(S) = S$ and hence $\{s_x\}$ is regular by Proposition 0.9. □

Proposition 0.29. If a Riemannian manifold $(M,g)$ admits a parallel s-structure then it is locally symmetric.

Proof. Let $\{s_x\}$ be a parallel s-structure on $(M,g)$, $\nabla S = 0$. Let $X,Y,Z \in M_p$ be tangent vectors and $\omega \in M_p^*$ a covector at $p \in M$. By parallel translation along each geodesic through $p$, $X,Y,Z,\omega$ can be extended to local vector fields $\tilde{X},\tilde{Y},\tilde{Z},\tilde{\omega}$ with vanishing covariant derivatives at $p$. Because $S$ is parallel, the local vector fields $S\tilde{X},S\tilde{Y},S\tilde{Z},S^{*-1}\tilde{\omega}$ have also vanishing covariant derivatives at $p$. (Here $S^*$ denotes the transpose map to $S$.) Now, because $R$ is invariant with respect to the affine transformations $s_x$, $x \in M$, we have

$$R(\omega,\tilde{X},\tilde{Y},\tilde{Z}) = R(S^{*-1}\tilde{\omega},S\tilde{X},S\tilde{Y},S\tilde{Z}), \qquad (29),$$

$$(\nabla R)(\omega,X,Y,Z;\ U) = (\nabla R)(S^{*-1}\omega,SX,SY,SZ;\ SU) \qquad (30).$$

Differentiating covariantly (29) in the direction of $SU$ at $p$ and using (30) we get $(\nabla R)(\omega,X,Y,Z;\ SU) = (\nabla R)(S^{*-1}\omega,SX,SY,SZ;\ SU) = (\nabla R)(\omega,X,Y,Z,U)$. Thus $(\nabla R)_p(\omega,X,Y,Z;(I-S)U) = 0$ for all $\omega \in M_p^*$, $X,Y,Z,U \in M_p$ and because $(I-S)_p$ is a non-singular transformation, we obtain $(\nabla R)_p = 0$. This holds for all $p \in M$ and hence $\nabla R = 0$. □

Due to Proposition 0.29, the regular s-structures constructed in our examples of g.s. spaces of higher order are all non-parallel. On the other hand, in both examples from 0.10 the regular s-structures are parallel. In particular, the geodesic symmetries of a Riemannian symmetric space form a parallel s-structure.

Let us consider now the second example - a family of rotations in the euclidean plane. We can generalize this example to higher dimensions as follows: Let $E^n$ be a euclidean space and $s$ an orthogonal transformation at the origin without fixed vectors. For each $x \in E^n$, let $t_x$ denote the translation such that $t_x(0) = x$. Then the family $\{s_x = t_x \circ s \circ t_x^{-1}, x \in M\}$ is a parallel regular s-structure. It is obvious that these families are the only parallel s-structures in $E^n$. Now, an interesting question is whether there are any non-parallel regular s-structures in euclidean spaces. In Note 4 we show that the answer is "no" for $E^2, E^3$ and $E^4$.

Surprisingly, for $E^5$ the answer is "yes". Let us identify $E^5$ with the space $C^2(z,w) \times R^1(t)$ and define a symmetry $\mathfrak{S}_o$ at the point $o = (0,0;\ 0)$ by the relations $z' = iw$, $w' = iz$, $t' = -t$. Further, we consider the following simply transitive group of isometries on $E^5$:

$$G:\quad z' = e^{it_o} z + z_o, \quad w' = e^{-it_o} w + w_o, \quad t' = t + t_o .$$

The underlying manifold for $G$ is $C^2(z_o, w_o) \times R^1(t_o)$. It is easy to see that the family $\{g \circ \mathfrak{S}_o \circ g^{-1} : g \in G\}$ is a regular s-structure of order 4 on $C^2 \times R$. We can show by a direct calculation that this s-structure is non-parallel. (See Note 4 for the details.)

## C a n o n i c a l   c o n n e c t i o n .

Let us recall that the Riemannian locally symmetric spaces are characterized by the relation $\nabla R = 0$. For generalized symmetric spaces we can hardly find a simple characterization in terms of the Riemannian curvature tensor. (One exception is the 3-symmetric case but some restrictions are still necessary.) Yet, if we are given a concrete regular s-structure $\{s_x\}$ in $(M,g)$, we are able to construct so called canonical connection of this s-structure. The canonical connection possesses just the simple properties we are looking for. It is natural to work with the local s-structures in this section. The basic formula (31) below is due to Graham and Ledger, [GL2].

<u>Definition 0.30.</u> A <u>local regular s-structure</u> on a Riemannian manifold $(M,g)$ is a family $\{s_x : x \in M\}$ of local isometries of

(M,g) such that

a) each local isometry $s_x$ has the point $x$ for an isolated fixed point (and it is then called a <u>local symmetry</u>),

b) the tangent tensor field $S$ defined by $S_x = (s_{x*})_x$ is smooth and invariant by all the local symmetries $s_x$.

<u>Proposition 0.31.</u> Let $\{s_x\}$ be a local regular s-structure on $(M,g)$ and let $\tilde{\nabla}$ be a connection on $M$ defined as follows:

$$\tilde{\nabla}_X Y = \nabla_X Y - D_X Y, \qquad D_X Y = (\nabla_{(I-S)^{-1}X} S)(S^{-1}Y) \qquad (31)$$

for all vector fields $X, Y \in \mathfrak{X}(M)$, where $\nabla$ is the Riemannian connection (and $D$ is a tensor field of type (1,2)). Then the connection $\tilde{\nabla}$ is invariant by all $s_x$ and it satisfies $\tilde{\nabla} S = 0$.

<u>Proof.</u> Because $\{s_x\}$ are local isometries, they leave invariant the Riemannian connection $\nabla$, and due to the regularity they preserve $S$. Hence we get successively the invariance of $\nabla S$, $D$ and $\tilde{\nabla}$. Now, we obtain the identity $(\tilde{\nabla}_X S)(Y) = 0$ by a direct calculation using the relations $S(D_X Y) = D_{SX} SY$, $D_{(I-S)X} SY = (\nabla_X S)(Y)$. □

<u>Proposition 0.32.</u> If $\nabla'$ is any connection invariant by all $s_x$ on $M$, and if $\tilde{\nabla}$ is the connection (31), then

$$\tilde{\nabla}_X Y = \nabla'_X Y - (\nabla'_{(I-S)^{-1}X} S)(S^{-1}Y) .$$

<u>Proof.</u> Let $E = \nabla' - \tilde{\nabla}$ be the corresponding difference tensor; we write $E_X Y = \nabla'_X Y - \tilde{\nabla}_X Y$ for $X, Y \in \mathfrak{X}(M)$. Since both $\nabla'$ and $\tilde{\nabla}$ are invariant under $s_x$, $x \in M$, then $E$ is invariant with respect to $S$: $S(E_X Y) = E_{SX} SY$ for $X, Y$ arbitrary. Now we check easily

$(E_{(I-S)^{-1}X} S)(S^{-1}Y) = E_{(I-S)^{-1}X} Y - S(E_{(I-S)^{-1}X} S^{-1}Y) =$

$= E_{(I-S)^{-1}X} Y - E_{S(I-S)^{-1}X} Y = E_X Y$. Since $\tilde{\nabla} S = 0$, we get finally $E_X Y = (E_{(I-S)^{-1}X} S)(S^{-1}Y) = (\nabla'_{(I-S)^{-1}X} S)(S^{-1}Y)$, which was to be proved. □

From the previous propositions we get

<u>Theorem 0.33.</u> Let $(M,g)$ be a Riemannian manifold and $\{s_x\}$ a local regular Riemannian s-structure on $M$. Then there is a unique connection $\tilde{\nabla}$ on $M$ such that

(i) $\tilde{\nabla}$ is invariant under all $s_x$,

(ii) $\tilde{\nabla} S = 0$.
In terms of the Riemannian connection, $\tilde{\nabla}$ is given by the formula (31).

Obviously, the connection $\tilde{\nabla}$ is fully determined by the local s-structure $\{s_x\}$, and the Riemannian connection plays only an indirect part here. Hence we can make the following definition:

Definition 0.34. The connection $\tilde{\nabla}$ from Theorem 0.33 is called the canonical connection of the local regular s-structure $\{s_x\}$.

Proposition 0.35. Let $\{s_x\}$ be a local regular s-structure on a manifold $(M,g)$. Then each tensor field invariant by all $s_x$ is parallel with respect to $\tilde{\nabla}$.

Proof. Because $\tilde{\nabla} S = 0$ and $s_x$ are local affine transformations with respect to $\tilde{\nabla}$, the proof is essentially the same as that of Proposition 0.29. We only have to replace the Riemannian connection $\nabla$ by the canonical connection $\tilde{\nabla}$ everywhere. The details are left to the reader. □

As a corollary we obtain

Proposition 0.36. Let $\{s_x\}$ be a local regular s-structure on a manifold $(M,g)$. Then with respect to the canonical connection we have $\tilde{\nabla} g = \tilde{\nabla} R = \tilde{\nabla} S = 0$, $\tilde{\nabla}(\nabla S) = \tilde{\nabla} D = 0$, $\tilde{\nabla} \tilde{R} = \tilde{\nabla} \tilde{T} = 0$.

Finally, we get

Proposition 0.37. Let $(M,g)$ be a Riemannian manifold admitting a local regular s-structure $\{s_x\}$. Then $(M,g)$ admits a subordinated analytic structure and the tensor field $S$ is analytic.

Proof. The connection $\tilde{\nabla}$ has parallel curvature and parallel torsion. Then with respect to the atlas consisting of normal coordinate systems, $M$ is an analytic manifold and the connection is analytic. (See Appendix B6.) Since $\tilde{\nabla} g = 0$, $\tilde{\nabla} S = 0$, the assertion follows. □

Now, in the global case we obtain

Proposition 0.38. Let $\{s_x\}$ be a regular s-structure on $(M,g)$. Then the canonical connection $\tilde{\nabla}$ is complete.

Proof. Let $p \in M$ be a fixed point. There exists $r > 0$ such that for every unit vector $X \in M_p$ the geodesic $\widetilde{\exp}\, sX$ with respect to $\widetilde{\nabla}$ is defined for $|s| \leq r$. Let $z_s = \widetilde{\exp}\, sZ$, $0 \leq s \leq a$, be any geodesic emanating from $p$ ($Z$ being a unit vector). Because $\widetilde{\nabla} g = 0$, the tangent vectors $\dot{z}_s$ along $z_s$ are all unit vectors. We shall show that $z_s$ can be extended to a geodesic defined for $0 \leq s \leq a + r$. Because the group $Cl(\{s_x\})$ is transitive on $M$, we can find a transformation $\varphi \in Cl(\{s_x\})$ which maps $p$ into $z_a$. Then $\varphi^{-1}$ is an isometry and it maps the unit vector $\dot{z}_a$ onto a unit vector $X$ at $p$. Since $\widetilde{\exp}\, sX$, $0 \leq s \leq r$, is a geodesic through $p$, and $\varphi$ is an affine transformation with respect to $\widetilde{\nabla}$, $\varphi(\widetilde{\exp}\, sX)$ is a geodesic through $z_a$. We put $z_{a+s} = \varphi(\widetilde{\exp}\, sX)$ for $0 \leq s \leq r$. Then $z_s$, $0 \leq s \leq a + r$, is a geodesic. This completes the proof. □

-------

It is well-known from the classical theory that a Riemannian symmetric space $(M,g)$ can be made a reductive homogeneous space: we only put $M = G/G_o$ where $G$ is the identity component of $I(M)$ and $G_o$ is an isotropy subgroup. Now, from 0.35 and 0.38 we can see that the canonical connection of a regular s-structure possesses similar properties as the canonical connection of a reductive homogeneous space (Cf. [KN II]). Hence we expect that a Riemannian manifold $(M,g)$ with a regular s-structure $\{s_x\}$ can be made a reductive homogeneous space in such a way that both canonical connections coincide.

In reality, we have more than one "natural" construction in the general case. We obtain the simplest one by putting $M = G/G_o$, where $G = Cl(\{s_x\})$, and $G_o$ is an isotropy subgroup. The fact that the homogeneous space $G/G_o$ is reductive, and also the coincidence of both canonical connections will be proved in Chapter II.

Now, in the previous definition of the canonical connection (Theorem 0.33), and in the last construction of the reductive homogeneous space $G/G_o$, our objects depended essentially on the regular s-structure $\{s_x\}$ whereas the Riemannian metric was only auxiliary. It is the aim of Chapters II and III that we get rid of Riemannian metrics and perform these constructions in a more general abstract situation. (This abstract situation leads to the concept of "generalized affine symmetric space".) The necessary theory of the reductive homogeneous spaces will be developed in Chapter I.

---

References: [GL2], [KN I-II], [K1], [K3], [K5], [K6], [KL], [LO], [TL2].

# CHAPTER I

## REDUCTIVE SPACES

For the standard concepts used in this Chapter see [KN I] and [Ch].

### Reductive homogeneous spaces.

Let $K$ be a connected Lie group and $H$ its closed subgroup. Consider the homogeneous manifold $K/H$. Here $\pi : K \to K/H$ will always denote the canonical projection, and $o = \pi(H)$ the origin of $K/H$.

Let $\underline{k} \supset \underline{h}$ be the Lie algebras of $K$ and $H$ respectively. Suppose that there is a subspace $\underline{m} \subset \underline{k}$ such that $\underline{k} = \underline{h} + \underline{m}$ (direct sum of vector spaces) and $ad(h)\underline{m} = \underline{m}$ for every $h \in H$. Then the homogeneous space $K/H$ is said to be <u>reductive with respect to the decomposition</u> $\underline{k} = \underline{h} + \underline{m}$. In this case we obviously have $[\underline{h},\underline{m}] \subset \underline{m}$.

<u>Proposition I.1.</u> Let $K/H$ be reductive with respect to a decomposition $\underline{k} = \underline{h} + \underline{m}$. Then the subspace $\underline{\ell} = \underline{m} + [\underline{m},\underline{m}]$ is an ideal of $\underline{k}$ and the corresponding connected normal subgroup $L \subset K$ is acting transitively on $K/H$ by the left translations. Moreover, $L$ is generated by the set $\exp(\underline{m})$, where $\exp: \underline{k} \to K$ is the exponential map at $e \in K$ (and $\underline{k}$ is naturally identified with the tangent space $K_e$).

<u>Proof.</u> It is obvious from the relation $[\underline{h},\underline{m}] \subset \underline{m}$ that $\underline{\ell}$ is an ideal of $\underline{k}$; thus the corresponding connected Lie subgroup $L \subset K$ is normal. Identifying $\underline{k}$ with $K_e$, we can see easily that $\pi_{*e}(\underline{\ell}) = \pi_{*e}(\underline{m}) = (K/H)_o$. Hence the transitivity of $L$ follows easily from the standard theorem on implicit functions and from the fact that $K/H$ is connected.

For the proof of the second statement, let $L'$ be the subgroup of $L$ generated by $\exp(\underline{m})$. Put $A = \{X \in \underline{\ell} \mid \exp tX \in L' \text{ for all } t \in R\}$. Let $\underline{a}$ be the subspace of $\underline{\ell}$ spanned by $A$. Then $\underline{m} \subset A \subset \underline{a}$. If $h \in L'$ and $X \in A$, we have $Ad(h)(\exp tX) = \exp(t \cdot ad(h)X) \in L'$, and hence $ad(h)X \in A$. It follows $ad(h)(\underline{a}) \subset \underline{a}$. In particular, for $Y \in A$, $X \in \underline{a}$, we have $e^{ad\, tY}X = ad(\exp tY) \cdot X \in \underline{a}$ for each $t$, and hence $[Y,X] \in \underline{a}$. By the linearity we get $[\underline{a},\underline{a}] \subset \underline{a}$; i.e., $\underline{a}$ is a subalgebra of $\underline{\ell}$. Now, $\underline{\ell} = \underline{m} + [\underline{m},\underline{m}] \subset \underline{a} + [\underline{a},\underline{a}] \subset \underline{a}$, i.e., $\underline{a} = \underline{\ell}$. Thus $A$ contains a basis $\{X_1,\dots,X_m\}$ of $\underline{\ell}$, and $\exp t_1X_1 \cdots$

$\dots \exp t_m X_m \in L'$ for all $t_i$. This shows that $L'$ contains a full neighborhood of $e$ in $L$ and therefore, it coincides with $L$. □

<u>Proposition I.2.</u> Let $K/H$ be reductive with respect to a decomposition $\underline{k} = \underline{h} + \underline{m}$. If $K$ acts on $K/H$ effectively, then the linear isotropy representation of $H$ in the tangent space $(K/H)_o$ is faithful.

<u>Proof.</u> Let $h \in H$, $h \neq e$, induce the identity transformation in $(K/H)_o$. Then $\mathrm{ad}(h)$ induces the identity transformation on $\underline{m}$. (We are using the fact that $\pi_{*e}: \underline{m} \longrightarrow (K/H)_o$ is an isomorphism.) Now, the automorphism $\mathrm{Ad}(h): K \longrightarrow K$ is the identity on $L$. Indeed, for each $X \in \underline{m}$ we have $\mathrm{Ad}(h)(\exp X) = \exp(\mathrm{ad}(h)X) = \exp X$, and by I.1, $\mathrm{Ad}(h)(g) = g$ for every $g \in L$. Further, because $L$ is transitive on $K/H$, we have an element $\ell \in L$ in each class $\bar{x} \in K/H$. Hence $h \cdot \bar{x} = h(\ell H) = \mathrm{Ad}(h)(\ell)H = \ell H = \bar{x}$. This means that $h$ induces the identity transformation on $K/H$, a contradiction to the effectivity of $K$. □

<u>Proposition I.3.</u> Let $K$ act effectively on the reductive homogeneous space $M = K/H$ ($\dim M = n$). Then the group $H$ is isomorphic to a subgroup $G$ of $GL(n,R)$, and the fibre bundle $K(M,H,\pi)$ is isomorphic to a subbundle $P(M,G)$ of the principal frame bundle $L(M,GL(n,R))$, (i.e., to a $G$-structure on $M$).

<u>Proof.</u> According to I.2, the isotropy representation of $H$ in $M_o$ is faithful. Hence the group $K$ acts freely on the frame bundle $L(M)$ (to the left); we shall denote this action by a dot.

Let $u_o$ be a fixed frame at $o$, and consider the map $f: K \longrightarrow L(M)$ given by the correspondence $g \longmapsto g \cdot u_o$. Further, let $\lambda: H \longrightarrow GL(n,R)$ denote the map given by the relation $h \cdot u_o = u_o \lambda(h)$ for each $h \in H$. Then $\lambda$ is injective and $f(gh) = f(g)\lambda(h)$ for $g \in K$, $h \in H$. (The map $\lambda$ is nothing but the isotropy representation of $H$ in $M_o$ calculated for a fixed coordinate system.) Consequently, if we put $G = \lambda(H)$, then $f$ determines an isomorphism of $K(M,H)$ onto a a subbundle $P(M,G) \subset L(M)$, as required. □

## The canonical connection.

<u>Convention I.4.</u> Let $K$ be a Lie group acting on a manifold $M$ to the left. Then we define a differentiable map $(\,\cdot\,): T(K) \times M \longrightarrow T(M)$

as follows: for $g \in K$, $X \in K_g$ and $p \in M$ we put

$$X \cdot p = \frac{d}{dt}\Big|_o (\exp_g tX \cdot p), \quad \text{where} \quad \exp_g = L_g \circ \exp \circ (L_{g^{-1}})_{*g} .$$

Let again $M = K/H$ be a reductive homogeneous space with respect to the decomposition $\underline{k} = \underline{h} + \underline{m}$ (we need not suppose that $K$ is effective on $M$). Let us consider the left-invariant distribution $\{\underline{m}_g, g \in K\}$ on $K$ generated by $\underline{m}$; here $\underline{m}_g \subset K_g$ for all $g \in K$. From the property $ad(h)\underline{m} = \underline{m}$ we see $\underline{m}_h = (L_h)_* \underline{m}_e = (R_h)_* \underline{m}_e$ for each $h \in H$. We shall usually identify $\underline{m}$ with $\underline{m}_e$.

Consider the frame bundle $L(M)$ with the structure group $GL(n,R)$ acting to the right, and the projection $\tilde{\pi}: L(M) \longrightarrow M$. The group $K$ acts on $L(M)$ to the left (where $g \cdot (e_1, \ldots, e_n) = (g_* e_1, \ldots, g_* e_n)$), and by I.4 we get a map $( \cdot ): T(K) \times L(M) \longrightarrow T(L(M))$.

<u>Lemma I.5.</u> The set of tangent subspaces $Q_u \subset (L(M))_u$ along the fibre $\tilde{\pi}^{-1}(o) \subset L(M)$ given by $Q_u = \underline{m} \cdot u$ is H-invariant, and it satisfies $\tilde{\pi}_*(Q_u) = M_o$, $Q_{us} = (R_s)_* Q_u$ for $u \in \tilde{\pi}^{-1}(o)$, $s \in GL(n,R)$.

<u>Proof.</u> For $h \in H$, $u \in \tilde{\pi}^{-1}(o)$ we have $(L_h)_* Q_u = (L_h)(\underline{m}_e \cdot u) = \underline{m}_h \cdot u =$
$= ((R_h)_* \underline{m}_e) \cdot u = \underline{m}_e \cdot (h \cdot u) = Q_{h \cdot u}$. Further, $\tilde{\pi}_*(Q_u) = \tilde{\pi}_*(\underline{m}_e \cdot u) = \underline{m}_e \cdot o =$
$= \pi_*(\underline{m}_e) = M_o$. Finally, we have $g \cdot (us) = (g \cdot u)s$ for $g \in K$, $u \in \tilde{\pi}^{-1}(o)$, $s \in GL(n,R)$. □

Because $K$ acts transitively on the set of all fibres of $L(M)$, we obtain:

<u>Proposition I.6.</u> There is a unique K-invariant connection in $L(M)$ such that the horizontal subspaces along the fibre $\tilde{\pi}^{-1}(o)$ are given by the rule $Q_u = \underline{m} \cdot u$ .

<u>Definition I.7.</u> The connection $\Gamma$ constructed in I.6 is called the <u>canonical connection</u> of the reductive homogeneous space $K/H$.

<u>Proposition I.8.</u> The canonical connection of the reductive homogeneous space $M = K/H$ is the unique K-invariant connection in $L(M)$ with the following property: for every frame $u$ at $o$, and for each $X \in \underline{m}$, the orbit $\exp(tX) \cdot u$ is horizontal.

<u>Proof.</u> We have $\frac{d}{dt}(\exp tX \cdot u) = \frac{d}{d\tau}\Big|_o (\exp(t+\tau)X \cdot u) =$
$= (\exp tX)_* \frac{d}{d\tau}\Big|_o (\exp \tau X \cdot u) = (\exp tX)_*(X \cdot u)$. Now, let $Q_z$ denote

the horizontal subspace of the canonical connection at $z \in L(M)$. Then we have $X \cdot u \in Q_u$ and $\frac{d}{dt}(\exp tX \cdot u) \in (\exp tX)_* Q_u = Q_{\exp tX \cdot u}$. Hence $\frac{d}{dt}(\exp tX \cdot u)$ is horizontal.

The uniqueness part is obvious from I.6. □

Corollary I.9. Consider the canonical connection $\Gamma$ of $M = K/H$.

(i) For each $X \in \underline{m}$ set $x(t) = \exp tX \cdot o$ in $M$. Then the parallel displacement of tangent vectors at $o$ along the curve $x(t)$, $0 \leq t \leq s$, coincides with the differential of $\exp sX \in K$ acting on $M$.

(ii) For each $X \in \underline{m}$, the curve $x(t) = \exp tX \cdot o$ is a geodesic. Conversely, every geodesic starting from $o$ is of the form $\exp tX \cdot o$ for some $X \in \underline{m}$.

(iii) The canonical connection on $M$ is complete.

Proof. (i) follows almost immediately from I.8. Indeed, choose $u \in L(M)$ at $o$. Since $\exp tX \cdot u$ is a horizontal curve which projects on the curve $x(t)$, $0 \leq t \leq s$, we can see that, for any $Y \in M_o$, $(\exp tX)_*(Y) = (\exp tX \cdot u) \cdot (u^{-1}Y)$ is parallel to $Y$ along the curve $x(t)$, $0 \leq t \leq s$. (Here $u^{-1}$ is considered as a map $u^{-1}: M_o \longrightarrow R^n$.) (ii) follows from (i) since the tangent vector $x^\cdot(t)$ is equal to $(\exp tX)_*(x^\cdot(0))$. (iii) is immediate from (ii) since $\exp tX$ is defined for all $t$. □

In accordance with Appendix A, we shall identify the canonical connection $\Gamma$ of $K/H$ and the corresponding affine connection $\nabla$ on $M$.

Proposition I.10. The canonical connection is the unique K-invariant affine connection on $M$ such that for every $X \in \underline{m}$ and every vector field $Y \in \mathfrak{X}(M)$ we have

$$(\nabla_{X^*}Y)_o = [X^*, Y]_o \qquad (1)$$

where $X^*$ denotes the vector field generated by the action of $X$ on $M$: $X^*_p = X \cdot p$ for each $p \in M$.

Proof. The infinitesimal version of (i), I.9, says that $(\nabla_{X^*}Y)_o = (\mathcal{L}_{X^*}Y)_o$, where $\mathcal{L}$ denotes the Lie derivative. Hence the canonical connection satisfies (1). On the other hand, the knowledge of the covariant derivatives $(\nabla_{X^*})_o$, $X \in \underline{m}$, and the K-invariance of the canonical connection determine $\nabla$ uniquely on $M$.

**Proposition I.11.** If a tensor field on M is invariant by K then it is parallel with respect to the canonical connection $\nabla$.

**Proof.** Let S be a K-invariant tensor field on M. Then $\mathcal{L}_{X^*}S = 0$ and hence $(\nabla_{X^*}S)_o = (\mathcal{L}_{X^*}S)_o = 0$ for every $X \in \underline{m}$, i.e., $(\nabla S)_o = 0$. (Here the covariant derivative $(\nabla_{X^*})_o$ and the Lie derivative $(\mathcal{L}_{X^*})_o$ are extended to derivations of the tensor algebra over $M_o$.) From the K-invariance of $\nabla$ and S we get $\nabla S = 0$ identically.

**Corollary I.12.** For the canonical connection of a reductive homogeneous space, the curvature tensor field and the torsion tensor field are parallel: $\nabla R = \nabla T = 0$.

**Proposition I.13.** For the curvature and torsion tensor fields of the canonical connection of M the following formulas hold:

$$T(X,Y)_o = -[X,Y]_{\underline{m}} \qquad \text{for } X,Y \in \underline{m} \qquad (2)$$

$$(R(X,Y)Z)_o = -[[X,Y]_{\underline{h}}, Z] \qquad \text{for } X,Y,Z \in \underline{m} \qquad (3)$$

where $\underline{m}$ is identified first with $\underline{m}_e \subset K_e$ and then with the tangent space $M_o$ by means of the projection $\pi: K \longrightarrow M$. (The indices at the brackets indicate taking the $\underline{m}$-component and the $\underline{h}$-component of a vector of $\underline{k}$.)

**Proof.** Strictly speaking, we have to prove the relations

$$T(X\cdot o, Y\cdot o) = -([X,Y]_{\underline{m}})\cdot o \qquad \text{for } X,Y \in \underline{m}, \qquad (2')$$

$$R(X\cdot o, Y\cdot o)(Z\cdot o) = -[[X,Y]_{\underline{h}}, Z]\cdot o \qquad \text{for } X,Y,Z \in \underline{m} \qquad (3').$$

A) Suppose first that K is acting effectively on M = K/H. Choose $u_o \in \tilde{\pi}^{-1}(o)$ in L(M). Consider the isomorphism $\lambda: H \longrightarrow G \subset GL(n,R)$ of Lie groups and the isomorphic imbedding f of K(M,H) into L(M,GL(n,R)) as a G-structure P(M,G) (see I.3). In particular, $f(g) = g\cdot u_o$ for each $g \in K$. We see easily that, for the canonical connection $\Gamma$ in L(M), the horizontal subspaces $Q_u$ at $u \in P(M,G)$ are always tangent to P(M,G). Thus the canonical connection possesses a reduction $\Gamma_P$ to the G-structure P(M,G).

Let X be a vector field on K, then the corresponding vector field $f_*(X)$ on P will be denoted by $X'$. Clearly, the map $X \longrightarrow X'$ is an isomorphism between the (infinite) Lie algebras $\mathfrak{X}(K)$ and $\mathfrak{X}(P)$. Put $\underline{m}' = \{X' \mid X \in \underline{m}\}$. Then $\underline{m}'$ consists of K-invariant and horizontal vector fields on $P$ (with respect to the connection $\Gamma_P$). Let us remark that for each $X \in \underline{m}$, $X'_{u_o} = X \cdot u_o \in P_{u_o}$ is a lift of $X \cdot o \in M_o$.

Now, choose $X, Y \in \underline{m}$, and consider the connection form $\omega$ and the canonical form $\theta$ (see Appendix A). Then $\omega(X') = \omega(Y') = 0$ because $X', Y'$ are horizontal, and $\theta(X'), \theta(Y') \in R^n$ are constant because $X', Y'$ are K-invariant on P. Then (A2) from Appendix A implies $\Theta(X',Y') = d\theta(X',Y') = \frac{1}{2}\{X'\theta(Y') - Y'\theta(X') - \theta([X',Y'])\} = -\frac{1}{2}\theta([X',Y'])$. Hence $T(X \cdot o, Y \cdot o) = u_o(2\Theta(X'_{u_o}, Y'_{u_o})) = -u_o(\theta[X',Y']_{u_o}) = -\widetilde{\pi}_*([X',Y']_{u_o})$. Here $[X',Y']_{u_o} = [X,Y]'_{u_o}$ is a lift of $[X,Y] \cdot o = [X,Y]_{\underline{m}} \cdot o$. Hence (2′) follows.

Further, for $X, Y \in \underline{m}$ we get from (A1) (Appendix A) $\Omega(X',Y') = d\omega(X',Y') = \frac{1}{2}\{X'\omega(Y') - Y'\omega(X') - \omega([X',Y'])\} = -\frac{1}{2}\omega([X',Y'])$. Choose $Z \in \underline{m}$ and put $\xi = u_o^{-1}(Z \cdot o)$. Then $R(X \cdot o, Y \cdot o)(Z \cdot o) = u_o(2\Omega(X'_{u_o}, Y'_{u_o}))\xi = -u_o(\omega([X',Y']_{u_o}))\xi = -u_o(\lambda_{*e}([X,Y]_{\underline{h}})\xi$. Recall that the linear isotropy representation of $H$ (or $\underline{h}$) in $M_o$ corresponds to the restriction to $\underline{m}$ of the adjoint representation of $H$ (or $\underline{h}$) in $\underline{k}$. (Here for $A \in \underline{h}$ and $w \in M_o$ the vector $Aw \in M_o$ is defined by $Aw = \frac{d}{dt}\Big|_o (\exp tA)_* w$.) Now, $[u_o \lambda(h)]\xi = h_*(u_o \xi) = h_*(Z \cdot o)$ for $h \in H$ according to the definition of $\lambda$, and hence $[u_o \lambda_*(A)]\xi = A(Z \cdot o)$ for $A \in \underline{h}$. Hence we obtain

$$-u_o(\lambda_*[X,Y]_{\underline{h}})\xi = -[X,Y]_{\underline{h}}(Z \cdot o) = -(\mathrm{ad}([X,Y]_{\underline{h}})Z) \cdot o .$$

This completes the proof.

B) Let us suppose now that $K$ is not effective on $K/H$ and denote by $N \subset K$ the closed normal subgroup of all elements $g \in K$ acting as the identity transformation on $M$. We have $N \subset H$; the Lie algebra $\underline{n}$ of $N$ is an ideal of $\underline{k}$, and $\underline{n} \subset \underline{h}$. Put $K' = K/N$, $H' = H/N$. Then the homogeneous space $K'/H'$ is canonically diffeomorphic to $M = K/H$, and $K'$ is acting on $M$ effectively. Also, $K'/H'$ is reductive with respect to the decomposition $\underline{k}' = \underline{h}' + \underline{m}$, where $\underline{k}' = \underline{k}/\underline{n}$, $\underline{h}' = \underline{h}/\underline{n}$. The action of the set $\exp(\underline{m})$ on $M$, or $L(M)$, has the same meaning for both homogeneous spaces $K/H$ and $K'/H'$ and a connection in $L(M)$ is $K'$-invariant if and only if it is K-invariant. According to Proposition I.8, the canonical connections of $K/H$ and $K'/H'$ coincide.

Now, (2′) and (3′) hold for the homogeneous space $K'/H'$. Denoting by $[\ ,\ ]$ and $\langle\ ,\ \rangle$ the Lie brackets in $\underline{k}$ and $\underline{k}'$ respectively, we get for $X, Y, Z \in \underline{m}$: $[X,Y]_{\underline{m}} = \langle X,Y\rangle_{\underline{m}}$, $[[X,Y]_{\underline{h}}, Z] \cdot o = \langle\langle X,Y\rangle_{\underline{h}'}, Z\rangle \cdot o$. Hence (2′) and (3′) hold for the homogeneous space K/H, too. □

Convention I.14. In the following, we shall always denote the canonical connection of a reductive homogeneous space by $\tilde{\nabla}$, and the corresponding curvature and torsion tensor field by $\tilde{R}$ and $\tilde{T}$ respectively.

Theorem I.15 (K.Nomizu,[N]). Let $K/H = M$ be a reductive homogeneous space with respect to a decomposition $\underline{k} = \underline{h} + \underline{m}$. Then there is a bijective correspondence between the K-invariant affine connections on M and between the linear maps $D: \underline{m}\times\underline{m} \to \underline{m}$ such that $D(ad(h)X, ad(h)Y) = ad(h)D(X,Y)$ for every $X,Y \in \underline{m}$, $h \in H$.

Proof. Let $\tilde{\nabla}$ denote the canonical affine connection in M and $\nabla$ a K-invariant affine connection. Then we identify $\underline{m}$ with $M_o$ and put $D = (\nabla - \tilde{\nabla})_o$ = the difference tensor at o.

Conversely, if D is a map of $\underline{m}\times\underline{m}$ into $\underline{m}$ as above, we put $\nabla_o = \tilde{\nabla}_o + D$. Then $\nabla_o$ is an H-invariant element of an affine connection at the origin o, and we can extend it in a unique way to a K-invariant affine connection $\nabla$ on M. (Here we take formula (A.10) from Appendix A into account.) □

## Algebraic characterization.

Proposition I.16. Let M be a manifold with an affine connection such that $\nabla T = \nabla R = 0$. Let $o \in M$ be arbitrary, and denote $V = M_o$. Then the following holds:

(a) For every $X,Y \in V$ the endomorphism $R_o(X,Y)$ acting as derivation on the tensor algebra $\mathcal{T}(V)$ satisfies
$R_o(X,Y)(R_o) = R_o(X,Y)(T_o) = 0$

(b) $R_o(X,Y) = -R_o(Y,X)$, $T_o(X,Y) = -T_o(Y,X)$

(c) (The first Bianchi identity)
$\mathfrak{S}\{R_o(X,Y)Z - T_o(T_o(X,Y),Z)\} = 0$ holds,

(d) (The second Bianchi identity)
$\mathfrak{S}\{R_o(T_o(X,Y),Z)\} = 0$ holds.

Here $\mathfrak{S}$ denotes the cyclic sum with respect to X,Y,Z.

Proof. (a) follows from the property $\nabla R = \nabla T = 0$, (b) is obvious. Finally, (c) and (d) are the Bianchi identities for the case $\nabla R = \nabla T = 0$ (see Appendix A, (A8) and (A9)). □

**Theorem I.17.** Let $V$ be a finite-dimensional real vector space, and $T_o, R_o$ tensors on $V$ of type $(1,2),(1,3)$ respectively satisfying the conditions (a) - (d). Then there is a unique connected and simply connected manifold $M$ with a complete affine connection $\tilde{\nabla}$ such that, for each point $x \in M$, there is a linear isomorphism $f: V \longrightarrow M_x$ satisfying $f(T_o) = \tilde{T}_x$, $f(R_o) = \tilde{R}_x$. (Here "unique" means exactly up to an affine diffeomorphism.) Moreover, $M$ can be made a reductive homogeneous space $K/H$ for which $\tilde{\nabla}$ is the canonical connection. In particular, we have $\tilde{\nabla}\tilde{T} = \tilde{\nabla}\tilde{R} = 0$.

Proof. Let $\underline{h}$ be the Lie algebra of all endomorphisms $A$ of $V$ which, as derivations of the tensor algebra $\mathcal{T}(V)$, satisfy $A(R_o) = A(T_o) = 0$. In particular, we have $R_o(X,Y) \in \underline{h}$ for every $X,Y \in V$ (axiom (a)). Let us define a Lie algebra $\underline{k}$ to be the direct sum $V + \underline{h}$ with the multiplication given by

$$\left.\begin{aligned} [X,Y] &= (-T_o(X,Y), -R_o(X,Y)) \\ [A,X] &= AX \\ [A,B] &= AB - BA \end{aligned}\right\} \quad X,Y \in V;\ A,B \in \underline{h} \qquad (4).$$

One can check easily that the Jacobi identities follow from the conditions (c) and (d). (Cf. K.Nomizu, [N].)

Let $K$ be the connected and simply connected Lie group with the Lie algebra $\underline{k}$ and let $H$ be the connected Lie subgroup corresponding to the Lie subalgebra $\underline{h} \subset \underline{k}$.

**Proposition I.18.** $H$ is a closed subgroup of $K$.

Proof. Consider the groups $ad_K(K) \supset ad_K(H)$ of inner automorphisms, and the Lie algebras $ad_{\underline{k}}(\underline{k}) \supset ad_{\underline{k}}(\underline{h})$ of inner derivations of $\underline{k} = V + \underline{h}$. Here $ad_{\underline{k}}(\underline{h})$ can be easily characterized as the subalgebra of all derivations of $\underline{k}$ which preserve $V$ and $\underline{h}$. Hence $ad_K(H)$ is the identity component of the group of all automorphisms of $\underline{k}$ which preserve $V$ and $\underline{h}$. Consequently, $ad_K(H)$ is a closed subgroup of the Lie group $Aut(\underline{k})$ of all automorphisms of $\underline{k}$ and also, it is a closed subgroup of $ad_K(K) = Int(\underline{k})$.

Now, the map $\underline{h} \longrightarrow ad_{\underline{k}}(\underline{h})$ is an isomorphism. If we denote $\tilde{\underline{h}} = \{X \in \underline{k} \mid ad(X) = ad(Y)$ for some $Y \in \underline{h}\}$, we get a direct sum decomposition $\tilde{\underline{h}} = \underline{c} \oplus \underline{h}$, where $\underline{c}$ is the centre of $\underline{k}$. Let $\tilde{H}, C$ be the connected subgroups of $K$ corresponding to $\tilde{\underline{h}}$ and $\underline{c}$. Because $ad_K(H)$ is closed in $ad_K(K)$, then $\tilde{H}$ is closed in $K$. On the other hand, $\tilde{H}$ is locally isomorphic to the direct product $H \times C$. Consequently, $H$ is closed in $\tilde{H}$ and thus in $K$. □

Proof of Theorem I.17 (continuation): Consider the homogeneous manifold $K/H$. It is simply connected and reductive with respect to the decomposition $\underline{k} = V + \underline{h}$. Denote by $\tilde{\nabla}$ the canonical connection of $K/H$.

Let us identify first $\underline{k} = V + \underline{h}$ with the tangent space $K_e$ and then $V$ with the tangent space $(K/H)_o$ at the origin via the projection $\pi : K \longrightarrow K/H$. Then $T_o, R_o$ considered as tensors in $(K/H)_o$ are invariant with respect to the isotropy representation of $H$ (see the definition of $\underline{h}$). Thus, we can extend $T_o, R_o$ to K-invariant tensor fields $T, R$ on $K/H$. Also, by Corollary I.9 we have $\tilde{\nabla} T = \tilde{\nabla} R = 0$. We show that $T$ and $R$ are equal to the torsion tensor field $\tilde{T}$ and the curvature tensor field $\tilde{R}$ of $\tilde{\nabla}$ respectively. In fact, according to (2), (3), (4) we have $\tilde{T}(X,Y)_o = -[X,Y]_V = T_o(X,Y)$, $(\tilde{R}(X,Y)Z)_o = -[[X,Y]_{\underline{h}}, Z] = R_o(X,Y)Z$ for $X,Y,Z \in V$. Because $T, \tilde{T}, R, \tilde{R}$ are K-invariant, we get $T = \tilde{T}$, $R = \tilde{R}$. For $x \in K/H$ let $g \in K$ be such that $\pi(g) = x$, and put $f = (L_g)_*$. Then $f(T_o) = \tilde{T}_x$, $f(R_o) = \tilde{R}_x$ as required. Finally, the affine manifold $(M, \tilde{\nabla})$ is unique according to Appendix B8. □□

The following result is very useful in applications:

Proposition I.19. The construction described in Theorem I.17 has the same outcome, i.e., it leads to the same manifold with an affine connection, $(M, \tilde{\nabla})$, if we replace the Lie algebra $\underline{h}$ by any subalgebra $\underline{h}' \subset \underline{h}$ supposing only that $R_o(X,Y) \in \underline{h}'$ for all $X, Y \in V$.

Proof. Let $M = K/H$ be the homogeneous space constructed in the proof of I.17. Then $\underline{k}' = V + \underline{h}'$ is a subalgebra of $\underline{k}$; let $K' \subset K$ be the corresponding connected subgroup. Then a standard argument shows that $K'$ acts transitively on $M$. Thus the subgroup $H' \subset H$ corresponding to $\underline{h}'$ is the maximal connected subgroup of $K'$ leaving the origin $o$ fixed. Hence, $H'$ is closed in $K'$.

Now, let $\tilde{K}'$ be the connected and simply connected Lie group with the Lie algebra $\underline{k}'$. Then we can consider $\tilde{K}'$ as the universal covering group of $K'$, and the connected subgroup $\tilde{H}' \subset \tilde{K}'$ corresponding to $\underline{h}'$ covers $H'$. Hence it follows that $\tilde{H}'$ is closed in $\tilde{K}'$. Further we proceed as in the second part of the proof of Theorem I.17.

<u>Corollary I.20.</u> Let $(M,\nabla)$ be a connected manifold with an affine connection such that $\nabla R = \nabla T = 0$. Then $(M,\nabla)$ is locally affinely diffeomorphic to a manifold $(\tilde{M},\tilde{\nabla})$, where $\tilde{M}$ is a simply connected reductive homogeneous space $K/H$ and $\tilde{\nabla}$ is its canonical connection. (Cf. [N].)

<u>Proof.</u> Use Theorem I.16, the proof of Theorem I.17 and Appendix B7.

## The group of transvections

Let $(M,\nabla)$ be a connected manifold with an affine connection. Let $u_o \in L(M)$ be a fixed frame at a point $o \in M$ and $P(u_o)$ the holonomy bundle through $u_o$, i.e., the set of all $u \in L(M)$ which can be joined to $u_o$ by a (piece-wise differentiable) horizontal curve. Further, let $\Phi(u_o)$ denote the holonomy group with reference frame $u_o$. ($\Phi(u_o) \subset GL(n,R)$ is isomorphic to the holonomy group $\Psi(o)$ with reference point $o$, cf. B5.) Let $\Gamma$ denote the corresponding connection in $L(M)$.

<u>I.21.</u> The famous <u>Reduction theorem</u> (see [KN I]) says that

(i) $P(u_o)$ is a differentiable subbundle of $L(M)$ with the structure group $\Phi(u_o)$.

(ii) The connection $\Gamma$ is reducible to a connection in $P(u_o)$.

Let $f: M \longrightarrow M$ be a diffeomorphism and $\tilde{f}: L(M) \longrightarrow L(M)$ the induced automorphism of $L(M)$. Clearly, if $\tilde{f}$ preserves a fixed holonomy bundle $P(u_o)$ then it also preserves the holonomy bundle $P(u)$ for each $u \in L(M)$.

<u>Definition I.22.</u> Let $(M,\nabla)$ be a connected manifold with an affine connection. The group of all affine transformations of $M$ preserving each holonomy bundle $P(u)$, $u \in L(M)$, is called <u>the group of transvections of $(M,\nabla)$</u>. It will be denoted by $Tr(M,\nabla)$, or simply by $Tr(M)$.

<u>Remark I.23.</u> More geometrically, an affine transformation $\Phi$ of

$(M,\nabla)$ belongs to the group $Tr(M)$ if and only if the following holds: for every point $p \in M$ there is a piece-wise differentiable curve $\gamma$ joining $p$ to $\Phi(p)$ such that the tangent map $\Phi_{*p}: M_p \longrightarrow M_{\Phi(p)}$ coincides with the parallel transport along $\gamma$.

Proposition I.24. The transvection group of $(M,\nabla)$ is a normal subgroup of the group $A(M)$ of all affine transformations.

Proof. Let $g \in Tr(M)$ and $\varphi \in A(M)$. Let $Q$ be a holonomy subbundle. Then $\tilde{g}(Q) \subset Q$. Further, $(\tilde{\varphi} \circ \tilde{g} \circ \tilde{\varphi}^{-1})(Q) = \tilde{\varphi} \circ \tilde{g}(\tilde{\varphi}^{-1}(Q)) \subset \tilde{\varphi}(\tilde{\varphi}^{-1}(Q)) = Q$. Here we used the fact that $\tilde{\varphi}^{-1}(Q)$ is a holonomy subbundle, too.

Theorem I.25. Let $(M,\nabla)$ be a connected manifold with an affine connection. Then the following two conditions are equivalent:

(i) The transvection group $Tr(M)$ acts transitively on each holonomy bundle $P(u) \subset L(M)$.

(ii) $M$ can be expressed as a reductive homogeneous space $K/H$ with respect to a decomposition $\underline{k} = \underline{m} + \underline{h}$, where $K$ is effective on $M$, and $\nabla$ is the canonical connection of $K/H$.

More specifically, if (i) is satisfied, then $Tr(M)$ is a connected Lie group and $M$ can be expressed in the form (ii) with $K = Tr(M)$. For every expression of $M$ in the form (ii), $Tr(M)$ is a normal Lie subgroup of $K$ and its Lie algebra is isomorphic to the ideal $\underline{\ell} = \underline{m} + [\underline{m}, \underline{m}]$ of $\underline{k}$. (Cf. also [KN II], p. 194.)

Remark I.26. The condition (i) means geometrically that to each piece-wise differentiable curve $\gamma$ joining some two points $p, q \in M$ there is a transformation $g \in Tr(M)$ such that $g(p) = q$, and the tangent map $g_{*p}: M_p \longrightarrow M_q$ coincides with the parallel transport along $\gamma$.

In particular, if (i) holds, then the isotropy subgroup of $Tr(M)$ at any point $p \in M$ is isomorphic to the holonomy group with reference point $p$.

Proof of Theorem I.25

(i) $\Longrightarrow$ (ii). Let $Tr(M)$ act transitively on a holonomy bundle $P(u_o)$, where $u_o \in L(M)$ is a frame at a point $o \in M$. Then $Tr(M)$ is simply transitive on $P(u_o)$. Let $A(M)$ be the Lie group of all affine transformations of $M$, and $A(u_o) \subset L(M)$ the subbundle generated by the action of $A(M)$ on $u_o$. The map $g \longmapsto g\cdot u_o$ is a diffeomorphism between $A(M)$ and $A(u_o)$.

According to I.21, $P(u_o)$ is a differentiable subbundle of $L(M)$ and hence that of $A(u_o)$. If we identify $Tr(M)$ with $P(u_o)$ and provide $Tr(M)$ with the corresponding differentiable structure, then $Tr(M)$ becomes a connected Lie subgroup of $A(M)$. In the following we shall denote $Tr(M)$ by $K$ and its Lie algebra by $\underline{k}$.

Let $\{Q_u\}$ be the horizontal distribution on $L(M)$ corresponding to the connection $\Gamma$. Because $\Gamma$ is reducible to $P(u_o)$, the subspaces $Q_u$ are tangent to $P(u_o)$ for $u \in P(u_o)$. Also, the distribution $\{Q_u\}$ is K-invariant.

Let $H \subset K$ be the subgroup preserving the fibre of $P(u_o)$ over $o$, then $M \approx K/H$ and $K$ is effective on $M$. Consider the linear isomorphism $\hat{u}_o\colon \underline{k} \longrightarrow (P(u_o))_{u_o}$ given by the map $X \longrightarrow X\cdot u_o$ for $X \in \underline{k}$. Then the decomposition of $(P(u_o))_{u_o}$ into the vertical subspace $\Phi_{u_o}$ and the horizontal subspace $Q_{u_o}$ corresponds to a decomposition $\underline{k} = \underline{h} + \underline{m}$, where $\underline{h}$ is the Lie algebra of $H$ and $\underline{m} \subset \underline{k}$ is a linear subspace. We obtain easily that $ad(H)\underline{m} \subset \underline{m}$. Moreover, $Q_u = \underline{m}\cdot u$ along the fibre of $L(M)$ over $o$. Thus $K/H$ is a reductive homogeneous space and $\Gamma$ is its canonical connection. □

-------

Suppose now that condition (ii) is satisfied and let $L$ denote the Lie group from I.1. We shall show successively: $L \subset Tr(M)$, $L$ is transitive on a bundle $P(u_o)$, and $L = Tr(M)$. Denote $\pi\colon K \longrightarrow K/H = M$.

**Lemma I.27.** For $p \in M$ denote by $\underline{m}_p$ the subspace $ad(g)\underline{m} \subset \underline{k}$, where $g \in K$ satisfies $g \in \pi^{-1}(p)$. Then $\underline{m}_p$ is independent of the choice of $g$ and $\underline{\ell} = \underline{m}_p + [\underline{m}_p, \underline{m}_p]$. Further, the group $L$ is generated by the set $\exp(\underline{m}_p)$.

**Proof.** The reductivity of $K/H$ means that $ad(h)\underline{m} = \underline{m}$ for $h \in H$; hence $\underline{m}_p$ is independent of $g \in \pi^{-1}(p)$. Further, $\underline{\ell}$ is an ideal of $\underline{k}$ and thus $ad(g)\underline{\ell} = \underline{\ell}$ for $g \in K$. It implies $\underline{m}_p + [\underline{m}_p, \underline{m}_p] = \underline{m} + [\underline{m}, \underline{m}] = \underline{\ell}$. Finally, $L$ is generated by $\exp(\underline{m})$ (cf. I.1) and hence $L = ad(g)L$ is generated by $ad(g)(\exp \underline{m}) = \exp(\underline{m}_p)$. □

**Lemma I.28.** The global 1-parametric subgroup $f_t = \exp tX$, $X \in \underline{m}_p$, has the property that the curve $x_t = f_t(p)$ is a geodesic and the tangent map $(f_s)_{*p}$ coincides with the parallel transport along the curve $x_t$, $0 \leqq t \leqq s$.

Proof. For $p = o$ (= the origin) the Lemma is equivalent with I.9, (i). For arbitrary $p \in M$ and $X \in \underline{m}_p$ put $X = ad(g)Y$, where $g \in K$, $Y \in \underline{m}$. Then $(\exp tY)(o)$ is a geodesic and the tangent map $(\exp sY)_{*o}$ coincides with the parallel transport along $(\exp tY)(o)$, $0 \leq t \leq s$. Now, $f_t = g \cdot \exp tY \cdot g^{-1}$ and $x_t = [g \cdot \exp tY](o)$. Because $g$ is an affine transformation of $(M, \nabla)$, then $x_t$ is a geodesic and $(f_s)_{*p} = g_* \circ (\exp sY)_{*o} \circ g_{*p}^{-1}$ coincides with the parallel transport along $x_t$, $0 \leq t \leq s$. □

Now, because $K$ is effective on $M = K/H$, we can consider $K$ as a group of affine transformations of $(M, \nabla)$. Denote by $Tr^*(M)$ the group of all transformations $g \in K$ with the following property: For each $p \in M$ there is a broken geodesic $\gamma$ from $p$ to $g(p)$ such that the transformation $g_{*p}$ coincides with the parallel transport along $\gamma$.

Obviously, $Tr^*(M) \subset Tr(M)$. We are going to prove $L \subset Tr^*(M)$. For each $p \in M$, let $Tr^*(p)$ denote the set of all transformations $g \in K$ such that $g_{*p}$ coincides with the parallel transport along a broken geodesic $\gamma$ from $p$ to $g(p)$. Then $Tr^*(M) = \bigcap_{p \in M} Tr^*(p)$. It suffices to show that $L \subset Tr^*(p)$ for each $p \in M$.

Let now $p \in M$ be given. According to Lemma I.27, $L$ is generated by the set $\exp(\underline{m}_p)$. This means that $L = \bigcup_{i=1}^{\infty} L^i(p)$, where

$$L^i(p) = \{g \in L \mid g = \exp X_1 \cdots \exp X_i;\ X_1, \ldots, X_i \in \underline{m}_p\}.$$

According to Lemma I.28 we have $L^1(p) \subset Tr^*(p)$. Suppose $L^i(p) \subset$ $\subset Tr^*(p)$ and let $g \in L^{i+1}(p)$, i.e., $g = g'h$ with $g' \in L^i(p)$ and $h = \exp X$, $X \in \underline{m}_p$. Then there is a broken geodesic $\gamma'$ from $p$ to $g'(p)$ such that $g'$ induces the parallel transport along $\gamma'$. On the other hand, we can write $g = h'g'$ where $h' = Ad(g')h = \exp X'$ for $X' = ad(g')X$. Here $X' \in \underline{m}_{g'(p)}$. Using Lemma I.28 once again for the new origin $g'(p)$ we see that the transformation $h'$ induces the parallel transport along a geodesic arc joining $g'(p)$ to $g(p) =$ $= (h'g')(p)$. Thus $g \in Tr^*(p)$. We have proved by induction that $L^i(p) \subset Tr^*(p)$ for all $i = 1, 2, \ldots$ and hence $L \subset Tr^*(M)$.

We shall need now some other lemmas.

Lemma I.29. For each broken geodesic $\gamma = pq$ in $M$ there is an element $g \in L$ such that the parallel transport along $\gamma$ coincides with the tangent map $g_{*p}$.

Proof. Let $\gamma$ consist of geodesic arcs $\gamma_1, \gamma_2, \dots, \gamma_k$ starting at the points $p = p_1, \dots, p_k$ respectively. Then the parallel transport along $\gamma_i$ is given by the differential of a transformation $h_i = \exp X_i$, where $X_i \in \underline{m}_{p_i} \subset \underline{\ell}$. Thus the parallel transport along $\gamma$ is given by the differential of $h = h_k \cdots h_1 \in L$. □

Lemma I.30. Let $L(p)$ be the isotropy group of $L$ at $p \in M$, and $L^{\bullet}(p)$ its connected component. Then $L^{\bullet}(p)$ is isomorphic to the restricted holonomy group $\Psi^{\bullet}(p)$ of $(M, \nabla)$ with the reference point $p$.

Proof. It is sufficient to prove our assertion for $p = o$. The Lie algebra of $L^{\bullet}(o)$ is isomorphic to $\underline{\ell} \cap \underline{h} = [\underline{m}, \underline{m}] \cap \underline{h}$. Thus it is spanned by all elements of the form $[X,Y]_{\underline{h}}$, $X, Y \in \underline{m}$. According to (3) and B9, it is isomorphic to the holonomy algebra $\underline{g}(o)$ with the reference point $o$. Hence the result follows. □

Now we shall prove that $L \subset Tr(M)$ acts transitively on any holonomy bundle $P(u_o)$, $u_o \in \tilde{\pi}^{-1}(o)$. According to the Reduction theorem I.21, if we consider the map $S \longrightarrow u_o S$ of $GL(n,R)$ onto the fibre $\tilde{\pi}^{-1}(o) \subset L(M)$, then the holonomy group $\Phi(u_o)$ is mapped onto $\tilde{\pi}^{-1}(o) \cap P(u_o)$ and the restricted holonomy group $\Phi^{\bullet}(u_o)$ is mapped onto the connected component $C(u_o)$ of $u_o$ in $\tilde{\pi}^{-1}(o) \cap P(u_o)$. Now, the group $L$ acts freely on $P(u_o)$, and for each $g \in L^{\bullet}(o)$ we have $g \cdot u_o \in C(u_o)$. Consider the homomorphism $\lambda : L^{\bullet}(o) \longrightarrow \Phi^{\bullet}(u_o)$ given by $g \cdot u_o = u_o \lambda(g)$. (Cf. the proof of Proposition I.3.) Because a) $\lambda$ is injective, b) $L^{\bullet}(o) \cong \Psi^{\bullet}(o) \cong \Phi^{\bullet}(u_o)$, c) $L^{\bullet}(o)$ and $\Phi^{\bullet}(u_o)$ are connected, we obtain $\lambda(L^{\bullet}(o)) = \Phi^{\bullet}(u_o)$. Consequently, $L^{\bullet}(o)$ is transitive on $C(u_o)$.

Now, each homotopy class of (piece-wise differentiable) loops at $o$ can be represented by a closed broken geodesic. According to Lemma I.29, the isotropy subgroup $L(o)$ is transitive on the fibre $\tilde{\pi}^{-1}(o) \cap P(u_o)$. Hence we obtain that $L$ is transitive on $P(u_o)$. As a consequence, we get the relation $L = Tr(M)$. This completes the proof of the Theorem. □ □

Remark I.31. For the implication (ii) $\longrightarrow$ (i), it is not essential that $K$ acts effectively on $M$. In fact, we have shown in Part B of the proof of Theorem I.13 that we can always replace $K$ by its effective factor group $K'$.

## A f f i n e   r e d u c t i v e   s p a c e s .

On the basis of the previous Theorem we can introduce the following definition:

Definition I.32. A connected manifold $(M,\nabla)$ with an affine connection is called an affine reductive space if the group $Tr(M,\nabla)$ acts transitively on each holonomy bundle $P(u_o) \subset L(M)$.

Thus, we have an "extrinsic" and an "intrinsic" definition of what we understand as a reductive space. Theorem I.25 also enables us to give a (1-1) correspondence between the objects described by our alternative definitions:

A reductive homogeneous space $K/H$ with the decomposition $\underline{k} = \underline{m} + \underline{h}$ is said to be prime if $K$ acts effectively on $K/H$ and $[\underline{m},\underline{m}]_{\underline{h}} = \underline{h}$. A connected manifold $(M,\nabla)$ with an affine connection is said to be pointed if there is given a fixed point (origin) $o \in M$.

Theorem I.33. There is a bijective correspondence between the pointed affine reductive spaces $(M,\widetilde{\nabla},o)$ and the prime reductive homogeneous spaces $K/H$ with a given decomposition. This correspondence is given by the formulas $K = Tr(M)$, $H = K_o$ in one direction, and by the formulas $M = K/H$, $\widetilde{\nabla}$ = canonical connection of $K/H$, in the other direction.

Proof: obvious.

Proposition I.34. Let $K$ be a connected Lie group and $\widetilde{\nabla}$ its canonical connection (i.e., the Cartan (-) connection). Then $(K,\widetilde{\nabla})$ is an affine reductive space and $Tr(K,\widetilde{\nabla}) = K_\ell$ = the group of all left translations of $K$.

Proof. Theorem I.25 says that, for $\underline{h} = 0$, the groups $Tr(K)$ and $K$ have isomorphic Lie algebras. On the other hand, $Tr(K)$ is a normal Lie subgroup of $K_\ell$. □

Theorem I.35. Let $(M_1,\widetilde{\nabla}_1)$, $(M_2,\widetilde{\nabla}_2)$ be affine reductive spaces. Then the direct product $(M_1 \times M_2,\ \widetilde{\nabla}_1 \times \widetilde{\nabla}_2)$ is an affine reductive space, and $Tr(M_1 \times M_2) = Tr(M_1) \times Tr(M_2)$.

Proof. The geometrical meaning of the product connection $\tilde{\nabla}_1 \times \tilde{\nabla}_2$ is the following: denote by $\pi_i : M_1 \times M_2 \longrightarrow M_i$ the projections; let $\gamma(t)$ be a path in $M_1 \times M_2$ and put $\gamma^i = \pi_i \circ \gamma$, $i = 1,2$. The parallel transport along $\gamma(\langle t_o, t_1 \rangle)$ is a linear map

$$\tau : ((M_1 \times M_2)_{\gamma(t_o)} \longrightarrow (M_1 \times M_2)_{\gamma(t_1)} .$$

Let $\tau^i : (M_i)_{\gamma^i(t_o)} \longrightarrow (M_i)_{\gamma^i(t_1)}$ be the parallel transport along the path $\gamma^i(\langle t_o, t_1 \rangle)$ in $M_i$, $i=1,2$. Using the canonical identification $(M_1 \times M_2)_{(x,y)} = (M_1)_x \oplus (M_2)_y$, we have

$$\tau = \tau^1 \oplus \tau^2 .$$

Let now $\alpha \in \mathrm{Tr}(M_1)$, $\beta \in \mathrm{Tr}(M_2)$, and let $(e_1, \dots, e_m, e_{m+1}, \dots, e_{m+n})$ be an adapted frame at a point $(x,y) \in M_1 \times M_2$. (It means that $e_1, \dots$ $\dots, e_m \in (M_1)_x$, $e_{m+1}, \dots, e_{m+n} \in (M_2)_y$.) Then $(\alpha \times \beta)_* (e_1, \dots, e_{m+n}) =$ $= (\alpha_*(e_1), \dots, \alpha_*(e_m), \beta_*(e_{m+1}), \dots, \beta_*(e_{m+n}))$. Here $(e_1, \dots, e_m)$ can be transformed into $\alpha_*(e_1, \dots, e_m)$ by a parallel transport in $M_1$, and $(e_{m+1}, \dots, e_{m+n})$ can be transformed into $\beta_*(e_{m+1}, \dots, e_{m+n})$ by a parallel transport in $M_2$. The synchronous parallel transport in $M_1 \times M_2$ sends $(e_1, \dots, e_{m+n})$ into $(\alpha \times \beta)_* (e_1, \dots, e_{m+n})$. Thus $\alpha \times \beta \in \mathrm{Tr}(M_1 \times M_2)$ and $\mathrm{Tr}(M_1) \times \mathrm{Tr}(M_2) \subset \mathrm{Tr}(M_1 \times M_2)$. By the similar argument we can prove that $\mathrm{Tr}(M_1) \times \mathrm{Tr}(M_2)$ acts transitively on each holonomy bundle of $(M_1 \times M_2, \tilde{\nabla}_1 \times \tilde{\nabla}_2)$ (cf. Remark I.26). Hence our space is affine reductive. The last observation also means that $\mathrm{Tr}(M_1) \times \mathrm{Tr}(M_2) = \mathrm{Tr}(M_1 \times M_2)$. □

Corollary I.36. Let $G_1, G_2$ be connected Lie groups and $\tilde{\nabla}_1, \tilde{\nabla}_2$ their canonical connections. Then $\tilde{\nabla}_1 \times \tilde{\nabla}_2$ is the canonical connection of $G_1 \times G_2$.

Proof. According to I.34, the transvection group of $(G_i, \tilde{\nabla}_i)$ can be identified with $G_i$, and the transvection group of the space $(G_1 \times G_2, \tilde{\nabla}_1 \times \tilde{\nabla}_2)$ can be identified with $G_1 \times G_2$. Let $\tilde{\nabla}$ be the canonical connection of $G_1 \times G_2$; then $G_1 \times G_2$ is also the transvection group of $(G_1 \times G_2, \tilde{\nabla})$. Because the action of $G_1 \times G_2$ on $G_1 \times G_2$ (to the left) is simply transitive, the parallelisms with respect to both connections $\tilde{\nabla}_1 \times \tilde{\nabla}_2$ and $\tilde{\nabla}$ are uniquely determined by this action, (and they are, in fact, independent of the path). Hence $\tilde{\nabla} = \tilde{\nabla}_1 \times \tilde{\nabla}_2$ . □

The converse of Theorem I.35 also holds:

Theorem I.37. Let $(M_1,\tilde{\nabla}_1),(M_2,\tilde{\nabla}_2)$ be affine manifolds such that $(M_1,\tilde{\nabla}_1)\times(M_2,\tilde{\nabla}_2) = (M_1\times M_2, \tilde{\nabla}_1\times\tilde{\nabla}_2)$ is an affine reductive space. Then $(M_1,\tilde{\nabla}_1),(M_2,\tilde{\nabla}_2)$ are affine reductive spaces.

Proof. By the parallel transport of an adapted frame along a path in $(M_1\times M_2, \tilde{\nabla}_1\times\tilde{\nabla}_2)$, we always get an adapted frame; namely $\tau(e_1,\dots,e_m, e_{m+1},\dots,e_{m+n}) = (\tau^1 e_1,\dots,\tau^1 e_m, \tau^2 e_{m+1},\dots,\tau^2 e_{m+n})$. Thus, for every $\gamma\in \mathrm{Tr}(M_1\times M_2)$ the image of an adapted frame is an adapted frame. It means that the group $\mathrm{Tr}(M_1\times M_2)$ preserves the tangent $M_1$-distribution and the tangent $M_2$-distribution; hence it preserves the families $\{M_1\times\{y\}, y\in M_2\}$, $\{\{x\}\times M_2, x\in M_1\}$. Choose a point $(x_0,y_0)\in M_1\times M_2$ and identify $M_1 = M_1\times\{y_0\}$, $M_2 = \{x_0\}\times M_2$. Put $G_1 = \{\gamma\in \mathrm{Tr}(M_1\times M_2)\,|\,\gamma(x_0,y_0)\in M_1\}$, $G_2 = \{\gamma\in \mathrm{Tr}(M_1\times M_2)\,|\,\gamma(x_0,y_0)\in M_2\}$. Then $G_i$ acts as a transitive transformation group of $M_i$, $i=1,2$. Obviously, $\mathrm{Tr}(M_1\times M_2) = G_1\times G_2$ and $G_i = \mathrm{Tr}(M_i)$, $i=1,2$. Hence the result follows. □

-------

Obviously, on an affine reductive space $(M,\tilde{\nabla})$, $\tilde{\nabla}$ is always complete and it satisfies $\tilde{\nabla}\tilde{R} = \tilde{\nabla}\tilde{T} = 0$. More generally, we have

Proposition I.38. On an affine reductive space $(M,\tilde{\nabla})$, a tensor field is parallel if, and only if, it is invariant with respect to the transvection group $\mathrm{Tr}(M)$.

Proof. It follows easily from the fact that $\mathrm{Tr}(M)$ acts transitively on each holonomy bundle $P(u)$, $u\in L(M)$. □

Proposition I.39. Let $(M,\tilde{\nabla})$ be a connected and simply connected affine reductive space, $\tilde{R}$ its curvature tensor field and $o\in M$ a fixed point. Then there is a bijective correspondence between the parallel tensor fields of type $(p,q)$ on $M$, and the tensors of type $(p,q)$ on $V = M_o$ which are annihilated by the derivations $\tilde{R}(X,Y)$, $X,Y\in V$, of the tensor algebra $\mathcal{T}(V)$. This correspondence is given by the evaluation of a tensor field at $o$.

Proof. Clearly the holonomy group $\Psi(o)$ coincides with the restricted holonomy group $\Psi^\bullet(o)$; hence it is connected. Thus $\Psi(o)$ is generated by the linear transformations $\exp X$, $X \in \underline{g}(o)$, where $\underline{g}(o)$ is the Lie algebra generated by all endomorphisms $\tilde{R}(X,Y)$, $X,Y \in M_o$ (cf. Appendix B9). Now, a tensor in $M_o$ is the evaluation of a parallel tensor field on $M$ if and only if it is invariant with respect to $\Psi(o)$. Hence the result follows. □

We shall close this Chapter with the following result (cf. [KN II], Theorem 2.8, p.194).

<u>Theorem I.40.</u> Let $(M,\tilde{\nabla})$ be a connected and simply connected manifold with a complete affine connection such that $\tilde{\nabla}\tilde{R} = \tilde{\nabla}\tilde{T} = 0$. Then $(M,\tilde{\nabla})$ is an affine reductive space.

<u>Proof.</u> Let $u_o \in L(M)$ be given and denote by $P(u_o)$ the corresponding holonomy bundle. Then for any $u_p \in P(u_o)$ let $\gamma$ be a piece-wise differentiable path from $o$ to $p$ such that $u_p$ is obtained from $u_o$ by the parallel transport $h_\gamma \colon M_o \longrightarrow M_p$ along $\gamma$. Because $\tilde{R}$ and $\tilde{T}$ are parallel, we obtain hence $h_\gamma(\tilde{R}_o) = \tilde{R}_p$, $h_\gamma(\tilde{T}_o) = \tilde{T}_p$. According to B8, there is an affine transformation $g$ of $M$ such that $g_{*o}$ coincides with $h_\gamma$. On the other hand, we can see easily that $\tilde{g}$ preserves the bundle $P(u_o)$. Thus $g \in Tr(M)$, and $Tr(M)$ is transitive on $P(u_o)$. □

---

References: [KN I-II],[K1],[K10],[KL],[Lo1],[Ch],[N].

# CHAPTER II

## DIFFERENTIABLE s-MANIFOLDS

### Affine symmetric spaces.

Let $M$ be a differentiable manifold with an affine connection $\nabla$. Using normal neighborhoods and exponential maps we can define local geodesic symmetries in the same way as we did in the Riemannian case.

Now, the manifold $(M,\nabla)$ is said to be affine locally symmetric if, for each $p \in M$, there is an open neighborhood $N_p$ such that the local geodesic symmetry in $N_p$ is an affine transformation. It is well-known that $(M,\nabla)$ is affine locally symmetric if and only if $T = 0$ and $\nabla R = 0$.

Further, a connected manifold $(M,\nabla)$ is said to be affine (globally) symmetric if each point $p \in M$ is an isolated fixed point of an involutive affine automorphism $s_p$ of $(M,\nabla)$. (Here $s_p$ is called a symmetry at $p$.) Obviously, a connected manifold $(M,\nabla)$ is affine symmetric if and only if it is affine locally symmetric and, for each point $p \in M$, a local geodesic symmetry with respect to $p$ can be extended to an affine automorphism $s_p$ of $(M,\nabla)$ onto itself. (The last condition is always satisfied if $M$ is connected and simply connected and $\nabla$ is complete.)

On an affine symmetric space $(M,\nabla)$, the system $\{s_p\}_{p\in M}$ of symmetries is uniquely determined, and for every two points $x,y \in M$ we have

$$s_x \circ s_y = s_z \circ s_x, \qquad \text{where} \quad z = s_y(y).$$

(The proof is the same as in the Riemannian case.) Also, the map $(x,y) \longrightarrow s_x(y)$ is differentiable. If we define a differentiable multiplication on $M$ by the formula $x \cdot y = s_x(y)$, we obtain a distributive groupoid - an object which is widely studied in the modern algebra.

Now, O.Loos [Lo1] has made the observation that the algebra of this distributive groupoid completely defines the geometry of an affine symmetric space. Following him, a symmetric space is a manifold $M$ with a differentiable multiplication $\mu: M \times M \longrightarrow M$ written as $\mu(x,y) = x \cdot y$ and with the following properties:

(1) $x \cdot x = x$

(2) $x \cdot (x \cdot y) = y$

(3) $x \cdot (y \cdot z) = (x \cdot y) \cdot (x \cdot z)$

(4) every $x$ has a neighborhood $U$ such that $x \cdot y = y$ implies $y = x$ for all $y$ in $U$.

For each $x \in M$, the map $s_x : M \longrightarrow M$ given by $s_x(y) = x \cdot y$ is called <u>the symmetry around $x$</u>. From (2) we get $(s_x)^2 = Id$ for each $x$, and consequently, each $s_x$ is a diffeomorphism.

Now, the main result on "symmetric spaces" is the following:

<u>Theorem II.1.</u> Each symmetric space $(M,\mu)$ admits a unique affine connection $\nabla$, called the canonical connection, which is invariant with respect to all symmetries $s_x$. $\nabla$ is complete and satisfies $T = 0$, $\nabla R = 0$. If, moreover, $M$ is connected, then the affine manifold $(M,\nabla)$ is the usual affine symmetric space and the symmetries $s_x$, $x \in M$, are the usual geodesic symmetries.

The main purpose of this Chapter is to characterize, in the same spirit, a wider class of spaces, which are "s-regular affine manifolds" in the sense of A.J.Ledger, [GL2]. Throughout this Chapter we shall make full use of the methods by O.Loos ([Lo1],[Lo2]). A great part of this theory was obtained independently by A.S.Fedenko and others (cf. [F1] - [F3]).

## The main theorem.

First of all, the axiom (4) says that the fixed point $x$ of $s_x$ is always isolated. Assuming (1) and (2), we can observe that (4) is equivalent to the following statement:

(4') The null vector of $T_x(M)$ is the unique fixed vector of the tangent map $(s_x)_{*x} : T_x(M) \longrightarrow T_x(M)$.

In fact, the axiom (2) implies that, for any point $x \in M$, there is a Riemannian metric $g_x$ on $M$ which is invariant under $s_x$. Hence $s_x$ is an isometry of $(M, g_x)$. Using the well-known formula $f \circ \exp = \exp \circ f_*$ for $f = s_x$, we obtain the wanted equivalence.

Now we can introduce our basic concept:

Definition II.2. A regular s-manifold is a manifold $M$ with a differentiable multiplication $\mu: M \times M \longrightarrow M$ (denoted by a dot) such that the maps $s_x: M \longrightarrow M$, $x \in M$, given by $s_x(y) = x \cdot y$ satisfy the following axioms:

(1) $s_x(x) = x$
(2′) each $s_x$ is a diffeomorphism
(3) $s_x \circ s_y = s_z \circ s_x$, where $z = s_x(y)$
(4′) for each $x \in M$, the tangent map $(s_x)_{*x}: T_x(M) \longrightarrow T_x(M)$ has no fixed vectors except the null vector.

We can see that (1),(3) have the same meaning as above whereas (2′) is a generalization of (2). The diffeomorphisms $s_x$, $x \in M$, are called symmetries of $M$, again.

Definition II.3. An automorphism of $(M,\mu)$ onto itself is a diffeomorphism $\Phi: M \longrightarrow M$ such that $\Phi(x \cdot y) = \Phi(x) \cdot \Phi(y)$ for every $x,y \in M$.

Obviously, all symmetries $s_x$ of $(M,\mu)$ are automorphisms due to axioms (3) and (2′). Now we shall state the main theorem of this chapter.

Theorem II.4. Let $(M,\mu)$ be a connected regular s-manifold with the symmetries $s_x$, $x \in M$. Denote by $S$ the tensor field of type (1,1) given by $S_x = (s_x)_{*x}$ for all $x \in M$. Then

(A) There is a unique connection $\tilde{\nabla}$ on $M$ (called the canonical connection) such that $\tilde{\nabla}$ is invariant under all $s_x$ and $\tilde{\nabla}S = 0$. $\tilde{\nabla}$ is complete and has parallel curvature and parallel torsion.

(B) The group $\mathrm{Aut}(M)$ of all automorphisms of $(M,\mu)$ is a transitive Lie transformation group, which is a closed subgroup of the full affine transformation group $A(M)$ with respect to $\tilde{\nabla}$.

(C) Let $G$ denote the identity component of $\mathrm{Aut}(M)$, let $o$ be a fixed point of $M$ and $G_o$ the corresponding isotropy group. Then the homogeneous space $G/G_o$ is reductive in a canonical way and, under the standard identification $G/G_o \cong M$, the connection $\tilde{\nabla}$ coincides with the canonical connection of $G/G_o$. (Cf. [K7].)

The proof is rather long and it will be given step by step in the next paragraphs.

## The canonical connection.

Let $(M,\mu)$ be a manifold with a multiplication (no other properties except the differentiability of the map $\mu: M\times M \longrightarrow M$ are required). Let $x,y\in M$, $u\in T_x(M)$, $v\in T_y(M)$. The point $\mu(x,y)$ is denoted by $x\cdot y$ as usual. Now, we define the products $x\cdot v$ and $u\cdot y$ in $T_{x\cdot y}(M)$ according to the rule

$$x\cdot v = \frac{d}{dt}\Big|_0 (x\cdot\alpha(t)), \quad u\cdot y = \frac{d}{dt}\Big|_0 (\beta(t)\cdot y) \tag{5}$$

where $\alpha(t)$, $\beta(t)$ are parametrized curves in $M$ such that $\frac{d\alpha(0)}{dt} = v$, $\frac{d\beta(0)}{dt} = u$.

Lemma II.5. Let $\alpha,\beta: (-\ell,\ell)\longrightarrow M$ be two parametrized curves in $M$, and let $\alpha\cdot\beta$ denote the curve given by $(\alpha\cdot\beta)(t) = \alpha(t)\cdot\beta(t)$, $t\in(-\ell,\ell)$. Then

$$\frac{d(\alpha\cdot\beta)(0)}{dt} = \frac{d\alpha(0)}{dt}\cdot\beta(0) + \alpha(0)\cdot\frac{d\beta(0)}{dt}.$$

Proof: See the "Leibniz formula", [KN I], p.10.

In accordance with Definition II.3, an automorphism of $(M,\mu)$ will be defined as a diffeomorphism $\Phi: M\longrightarrow M$ such that $\Phi(x\cdot y) = \Phi(x)\cdot\Phi(y)$, $x,y\in M$. We shall often denote the tangent map of $\Phi$ by the same symbol $\Phi$.

Lemma II.6. Let $\Phi$ be an automorphism of $(M,\mu)$ and $x\in M$, $u\in T(M)$. Then

$$\Phi(x\cdot u) = \Phi(x)\cdot\Phi(u), \quad \Phi(u\cdot x) = \Phi(u)\cdot\Phi(x) \tag{6}$$

Proof: obvious.

As usual, $\mathfrak{X}(M)$ will denote the Lie algebra of all differentiable vector fields on $M$. The elements of $\mathfrak{X}(M)$ are considered as the cross-sections $M\longrightarrow T(M)$.

Definition II.7. (Cf. [Lo1], p.51.) Let $(M,\mu)$ be a manifold with multiplication. A derivation of $(M,\mu)$ is a vector field $X\in\mathfrak{X}(M)$ such that

$$X(p\cdot q) = X(p)\cdot q + p\cdot X(q) \quad \text{for all} \quad p,q \in M.$$

The set of all derivations of $(M,\mu)$ will be denoted by $\mathrm{Der}(M)$.

Proposition II.8. a) The derivations of $(M,\mu)$ form a Lie subalgebra of the Lie algebra $\mathcal{X}(M)$.

b) A one-parameter group of transformations of $M$ is a group of automorphisms if and only if the corresponding vector field is a derivation.

Proof. a) follows by the direct calculation of a Lie bracket $[X,Y]$ at the point $p\cdot q$. Here we use the obvious relations $(u\cdot x)(f) = u(f\circ R_x)$, $(y\cdot v)(f) = v(f\circ L_y)$ for $x,y \in M$, $u,v \in T(M)$, $f \in \mathcal{F}(M)$.

b) Let $\{\Phi_t\}$ be the given one-parameter group. Then $\{\Phi_t\}$ is a group of automorphisms if and only if $\Phi_t(p\cdot q) = \Phi_t(p)\cdot\Phi_t(q)$. Now, the result follows easily by considering the curves $\alpha(t) = \Phi_t(p)$, $\beta(t) = \Phi_t(q)$ in Lemma II.5. □

In the future, $(M,\mu)$ will always denote a regular s-manifold. As we know, each symmetry $s_x$ is an automorphism of $(M,\mu)$ and so is its inverse $s_x^{-1}$. To get a unified symbolism, we shall write simply $x^{-1}\cdot y$ instead of $s_x^{-1}(y)$ for $x,y \in M$.

Notice that, due to Formula (5), we have

$$x\cdot u = (s_x)_* u \quad \text{for every} \quad x \in M, \quad u \in T(M). \qquad (7).$$

Lemma II.9. Let $p \in M$, $v \in T_p(M)$, $x,y \in M$; then

$$v = p\cdot v + v\cdot p \qquad (8)$$

$$v\cdot(x\cdot y) = (v\cdot x)\cdot(p\cdot y) + (p\cdot x)\cdot(v\cdot y). \qquad (9).$$

Proof. Choose a curve $\gamma\colon (-\ell,\ell) \longrightarrow M$ such that $\gamma(0) = p$, $\frac{d\gamma(0)}{dt} = v$, and put $\alpha(t) = \beta(t) = \gamma(t)$ for each $t$. Then we obtain (8) from the "Leibniz formula" (Lemma II.5) using the axiom $x\cdot x = x$. Further, put $\alpha(t) = \gamma(t)\cdot x$, $\beta(t) = \gamma(t)\cdot y$ for each $t$, then $(\alpha\cdot\beta)(t) = \gamma(t)\cdot(x\cdot y)$ and using Leibniz' formula we obtain (9). □

Consider now the tensor field $S$ on $M$ defined by the formula $S_x = (s_x)_{*x}$, $x \in M$. According to axioms (2') and (4') of Definition II.2, we can construct also the tensor fields $S^{-1}$, $I-S$, $(I-S)^{-1}$.

Recall that a diffeomorphism $\Phi$ of M induces an automorphism $\tilde{\Phi}$ of the algebra $\mathcal{T}(M)$ of all global tensor fields on M (cf. [KN I]).

<u>Proposition II.10.</u> If $\Phi$ is an automorphism of $(M,\mu)$ then $\tilde{\Phi}$ preserves the tensor fields $S$, $S^{-1}$, $I-S$, $(I-S)^{-1}$.

<u>Proof.</u> Let $p \in M$, $u \in T_p(M)$; then $\Phi(S_p u) = \Phi(p \cdot u) = \Phi(p)\cdot\Phi(u) = S_{\Phi(p)}\Phi(u)$. Thus $\Phi \circ S_p = S_{\Phi(p)} \circ \Phi$ on $T_p(M)$ and $\tilde{\Phi}$ preserves the tensor field $S$. The other cases are now easy. □

<u>Proposition II.11.</u> Let $v$ be an arbitrary tangent vector of M and $p \in M$ its initial point. Define a map $L(v)\colon M \longrightarrow T(M)$ by the formula

$$L(v)(x) = (I_p - S_p)^{-1}v\cdot(p^{-1}\cdot x), \qquad x \in M \qquad (10).$$

Then $L(v)$ is a derivation of $(M,\mu)$ and $L(v)(p) = v$. The map $L\colon T(M) \longrightarrow Der(M)$ is linear and injective on each tangent space $T_p(M)$.

<u>Proof.</u> We can see easily that $L(v)(x) \in T_x(M)$ for each $x$, and thus $L(v) \in \mathfrak{X}(M)$. Using Formula (9) we get $L(v)(x\cdot y) =$
$= (I_p - S_p)^{-1}v\cdot[p^{-1}\cdot(x\cdot y)] = (I_p - S_p)^{-1}v\cdot[(p^{-1}\cdot x)\cdot(p^{-1}\cdot y)] =$
$= [(I_p - S_p)^{-1}v\cdot(p^{-1}\cdot x)]\cdot[p\cdot(p^{-1}\cdot y)] + [p\cdot(p^{-1}\cdot x)]\cdot[(I_p - S_p)^{-1}v\cdot(p^{-1}\cdot y)] =$
$= L(v)(x)\cdot y + x\cdot L(v)(y)$, which proves the first statement. Further, using Formula (8) we get $L(v)(p) = (I_p - S_p)^{-1}v\cdot(p^{-1}\cdot p) =$
$= (I_p - S_p)^{-1}v\cdot p = (I_p - S_p)^{-1}v - p\cdot(I_p - S_p)^{-1}v = I_p(I_p - S_p)^{-1}v -$
$- S_p(I_p - S_p)^{-1}v = v$. The last assertion is now obvious. □

<u>Proposition II.12.</u> For every automorphism $\Phi$ of $(M,\mu)$ we have $\Phi_* \circ L = L \circ \Phi_*$ on $T(M)$, with the values in $Der(M)$.

<u>Proof.</u> Choose $p, x \in M$, $v \in T_p(M)$. Using (6) and Proposition II.10 we obtain $L(\Phi_* v)(x) = (I_{\Phi(p)} - S_{\Phi(p)})^{-1}(\Phi_* v)\cdot([\Phi(p)]^{-1}\cdot x) =$
$= \Phi_*((I_p - S_p)^{-1}v)\cdot([\Phi(p)]^{-1}\cdot x) = \Phi_*((I_p - S_p)^{-1}v)\cdot\Phi(p^{-1}\cdot\Phi^{-1}(x)) =$
$= \Phi_*[L(v)(\Phi^{-1}(x))] = (\Phi_* L(v))(x)$. □

Proposition II.13. The formula

$$\tilde{\nabla}_v Y = [L(v), Y](p) \qquad (p \in M,\ v \in T_p(M),\ Y \in \mathcal{X}(M)) \qquad (11)$$

defines an affine connection $\tilde{\nabla}$ on M. Each automorphism of $(M,\mu)$ is an affine transformation of $(M,\tilde{\nabla})$, and each derivation of $(M,\mu)$ is an infinitesimal affine transformation of $(M,\tilde{\nabla})$.

Proof. Obviously, the above operation is linear with respect to v and Y. Now, $\tilde{\nabla}_v(fY) = [L(v),fY](p) = (L(v)(p))(f)\cdot Y(p) + f(p)\cdot[L(v),Y](p) = v(f)\cdot Y(p) + f(p)\cdot\tilde{\nabla}_v Y$. (Here the dots denote the usual multiplication of vectors by real numbers.)

Further, let $\Phi \in \mathrm{Aut}(M)$. Then using Proposition II.12 we obtain $\tilde{\nabla}_{\Phi_*(v)}\Phi_*(Y) = [L(\Phi_*(v)),\Phi_* Y](\Phi(p)) = (\Phi_*[L(v),Y])(\Phi(p)) = \Phi_*([L(v),Y](p)) = \Phi_*(\tilde{\nabla}_v Y)$.

Finally, let $X \in \mathrm{Der}(M)$. Then X generates a local group of local automorphisms of $(M,\mu)$ (an easy modification of Proposition II.8) and thus a local group of local affine transformations of $(M,\mu)$. Consequently, X is an infinitesimal affine transformation. □

Corollary II.14. The connection $\tilde{\nabla}$ is invariant under all $s_x$, $x \in M$.

Proposition II.15. $\tilde{\nabla} S = 0$.

Proof. For $v \in T_p(M)$, $Y \in \mathcal{X}(M)$ we have $\tilde{\nabla}_v Y = (\mathcal{L}_{L(v)}Y)(p)$. Now, $(\tilde{\nabla}_v S)(Y) = \tilde{\nabla}_v(SY) - S(\tilde{\nabla}_v Y) = [\mathcal{L}_{L(v)}SY - S(\mathcal{L}_{L(v)}Y)](p) = [(\mathcal{L}_{L(v)}S)(Y)](p)$. On the other hand, $L(v)$ is an infinitesimal automorphism, and according to the proof of Proposition II.10, S is invariant with respect to the local automorphisms of $(M,\mu)$. Hence $\mathcal{L}_{L(v)}S = 0$, and the Proposition follows. □

Proposition II.16. ("Ledger's formula.") Let $\nabla$ be an $s_x$-invariant connection on $(M,\mu)$. Then

$$\tilde{\nabla}_X Y = \nabla_X Y - (\nabla_{(I-S)^{-1}X} S)(S^{-1}Y) \qquad (12)$$

for all vector fields $X,Y \in \mathcal{X}(M)$.

Proof: see the proof of Proposition 0.32.

Summarizing II.14-16 we obtain

Theorem II.17. For each regular s-manifold $(M,\mu)$ (not necessarily connected) there is a unique connection $\tilde{\nabla}$ which is invariant with respect to all $s_x$ and such that $\tilde{\nabla} S = 0$. It is given by Formula (11), II.13.

Definition II.18. The connection $\tilde{\nabla}$ given by (11) will be called the canonical connection of $(M,\mu)$.

Convention II.19. We shall use an alternative symbol $(M,\{s_x\})$ for a regular s-manifold $(M,\mu)$ in the future.

## Regular homogeneous s-manifolds.

From now on we shall suppose the manifold M to be connected. Let us denote by $A(M)$ the Lie group of all affine transformations of M with respect to the connection $\tilde{\nabla}$.

Proposition II.20. A transformation $\Phi \in A(M)$ is an automorphism of $(M,\{s_x\})$ if and only if it preserves the tensor field S. Consequently, $Aut(M)$ is a closed subgroup of $A(M)$ and hence a Lie transformation group of M.

Proof. $\Phi$ preserves the tensor field S if and only if, for each $x \in M$, the maps $\Phi \circ s_x$, $s_{\Phi(x)} \circ \Phi$ have the same tangent map at x. Further, $\Phi$ is an automorphism if and only if, for each x, the maps $\Phi \circ s_x$, $s_{\Phi(x)} \circ \Phi$ coincide. Because $\Phi \circ s_x$, $s_{\Phi(x)} \circ \Phi$ are affine transformations, both conditions are equivalent to each other. □

Proposition II.21. The Lie transformation group $Aut(M)$ acts transitively on M.

Proof. (Cf. [LO].) Choose an origin $o \in M$ and let $K \subset Aut(M)$ be the transformation group of M generated (algebraicaly) by all the symmetries $s_x$, $x \in M$. We shall prove that the orbit $K^{(o)}$ of o with respect to K is an open submanifold of M, and thus $K^{(o)} = M$.

Consider the map $\lambda(x) = s_x(p) = x \cdot p$, where p belongs to $K^{(o)}$ and x runs over M. We have $\lambda(p) = p$. For $v \in T_p(M)$ we get $\lambda_{*p}(v) = v \cdot p = v - p \cdot v = (I_p - S_p)v$ (see Formula (8)). Hence $\lambda_{*p} = I_p - S_p$ is a non-singular transformation and $\lambda$ maps a neighborhood U of p diffeomorphically onto a neighborhood V of p. We get $V \subset K^{(o)}$ and the orbit $K^{(o)}$ is open. The union of all other orbits of K must be also open and hence $K^{(o)}$ is closed. Consequently, $K^{(o)} = M$. □

Let us recall the well-known "Fitting's Lemma" (see [J], Ch. II).

Lemma II.22. Let W be a finite dimensional vector space and $A: W \to W$ an endomorphism. Then there is a unique decomposition $W = W_{0A} \oplus W_{1A}$ of W into A-invariant subspaces (called the Fitting 0-component and the Fitting 1-component respectively) such that the restriction of A to $W_{0A}$ is nilpotent and the restriction of A to $W_{1A}$ is an automorphism. Further, we have

$$W_{1A} = \bigcap_{i=1}^{\infty} A^i(W), \qquad W_{0A} = \{v \mid A^i(v) = 0 \quad \text{for some} \quad i\} \quad (13).$$

The following definition is due to N.A.Stepanov, [S2].

Definition II.23. A regular homogeneous s-manifold is a triplet $(G,H,\sigma)$, where G is a connected Lie group, H its closed subgroup and $\sigma$ an automorphism of G such that the following two axioms are satisfied:

(i) $(G^{\sigma})^{\bullet} \subset H \subset G^{\sigma}$, where $G^{\sigma} = \{g \in G \mid \sigma(g) = g\}$, $(G^{\sigma})^{\bullet} =$ = the identity component of $G^{\sigma}$.

(ii) If $\sigma_*$ denotes the induced automorphism of the Lie algebra $\underline{g}$ of G, and A denotes the (linear) endomorphism $id - \sigma_*$, then $\underline{g}_{0A} = \underline{h}$ (= the Lie algebra of H).

Proposition II.24. If $(G,H,\sigma)$ is a regular homogeneous s-manifold, then

(a) $\underline{h} = \operatorname{Ker} A$, $\underline{g}_{1A} = \operatorname{Im} A$,

(b) the homogeneous space G/H is reductive with respect to the decomposition $\underline{g} = \underline{h} + \underline{g}_{1A}$.

Proof. (a) follows from the relations $\sigma_*|_{\underline{h}} = id$, $\underline{h} = \underline{g}_{0A}$ and from II.22. Further, because $H \subset G^{\sigma}$, we obtain $ad(h) \circ \sigma_* = \sigma_* \circ ad(h)$ for each $h \in H$. Hence $ad(h)$ commutes with A on $\underline{g}$ and $ad(h)(\underline{g}_{1A}) = (ad(h) \circ A)(\underline{g}) = A(ad(h)\underline{g}) = A(\underline{g}) = \underline{g}_{1A}$. □

A great part of the following theorem is due to Stepanov and Fedenko. (See [F1],[F3],[S2].)

Theorem II.25.

I) Let $(G,H,\sigma)$ be a regular homogeneous s-manifold, $\pi : G \to G/H$ the canonical projection, and let $s$ be the transformation of $G/H$ induced by $\sigma$, i.e., we have $s\circ\pi = \pi\circ\sigma$ on $G$.

Let $G$ act on $G/H$ by the left translations. Then putting

$$s_{\pi(g)} = g\circ s\circ g^{-1} \quad \text{for each} \quad g \in G \qquad (14),$$

we obtain a well-defined family $\{s_x\}_{x\in G/H}$ of diffeomorphisms of $G/H$. $(G/H,\{s_x\})$ is a connected regular s-manifold and $G$ acts as a group of automorphisms of it.

II) Conversely, let $(M,\{s_x\})$ be a connected regular s-manifold and $o \in M$ a fixed point. Let $G$ be the identity component of the automorphism group $\mathrm{Aut}(M)$ and $G_o$ the isotropy subgroup of $G$ at $o$. Define a map $\sigma : G \to \mathrm{Aut}(M)$ by the formula

$$\sigma(g) = s_o\circ g\circ s_o^{-1}, \quad g \in G \qquad (15).$$

Then $\sigma$ is an automorphism of $G$, and $(G,G_o,\sigma)$ is a regular homogeneous s-manifold. Also, $M \approx G/G_o$ and the symmetries $s_x$ are given by the formula (14) where $s = s_o$.

III) Finally, in both cases I and II, the canonical connection of the regular s-manifold coincides with the canonical connection of the corresponding reductive homogeneous space.

Proof. Ad I): We shall identify the elements of $G$ with the corresponding transformations of $M = G/H$. (Here we need not suppose that $G$ is effective.) Choose $g \in G$ and $x \in M$, then $x = \pi(g')$ for some $g' \in G$. Now, $(s\circ g\circ s^{-1})(x) = (s\circ g\circ s^{-1}\circ\pi)(g') = (s\circ g\circ\pi)(\sigma^{-1}(g')) = (s\circ\pi)[g\sigma^{-1}(g')] = (\pi\circ\sigma)(g\sigma^{-1}(g')) = \pi(\sigma(g)g') = \sigma(g)[\pi(g')] = \sigma(g)(x)$. Hence we get

$$s\circ g\circ s^{-1} = \sigma(g) \quad \text{for each} \quad g \in G \qquad (16).$$

In particular, for $h \in H$ we obtain $s\circ h\circ s^{-1} = h$ and hence $h\circ s\circ h^{-1} = s$. Consequently, the transformation $g\circ s\circ g^{-1}$ always depends only on $\pi(g)$ and (14) defines a family $\{s_x\}_{x\in M}$ of diffeomorphisms of $M$.

Using local sections of the bundle $\pi : G \to M$ we can see easily that the map $(x,y) \mapsto s_x(y)$ is differentiable. Further, for $x \in M$, $x = \pi(g)$, we have $x = g(o)$ ($o$ = the origin) and hence $s_x(x) = (g\circ s\circ g^{-1})(x) = x$, because $s(o) = o$. Now, for $x,y \in M$ put $s_x = g\circ s\circ g^{-1}$, $s_y = g'\circ s\circ (g')^{-1}$, where $x = g(o)$ and $y = g'(o)$. Then

$(g \circ s \circ g^{-1} \circ g' \circ s^{-1})(o) = s_x(g'(o)) = s_x(y)$; on the other hand, (16) yields $g \circ s \circ g^{-1} \circ g' \circ s^{-1} = g\,\sigma(g^{-1}g')$. Thus, the map $g \circ s \circ g^{-1} \circ g' \circ s^{-1}$ coincides with the action of an element $g'' \in G$, $g''(o) = s_x(y)$. Now, $s_x \circ s_y = g \circ s \circ g^{-1} \circ g' \circ s \circ (g')^{-1} = g'' \circ s \circ (g'')^{-1} \circ g \circ s \circ g^{-1} = s_{s_x(y)} \circ s_x$. It remains to prove axiom (4′). Obviously, if we identify $\underline{g}$ with $T_e(G)$, then the projection $\pi_{*e}: T_e(G) \longrightarrow T_o(M)$ induces an isomorphism of $\underline{g}_{1A}$ onto $T_o(M)$ (Proposition II.24). From the relation $\pi_* \circ \sigma_* = s_* \circ \pi_*$ we can see that $\pi_* \circ A = (I_o - s_{*o}) \circ \pi_*$. Because $A$ is an automorphism on $\underline{g}_{1A}$, $I_o - s_{*o}$ is an automorphism of $T_o(M)$. From (14) we obtain easily that $I_p - S_p$ is an automorphism of $T_p(M)$ for each $p \in M$. Thus $(M, \{s_x\})$ is a regular s-manifold.

Finally, let $g \in G$ and $x \in M$ be given; we can suppose $x = g'(o)$ for some $g' \in G$. Then $g \circ s_x = g \circ g' \circ s \circ (g')^{-1} = (g \circ g') \circ s \circ (g \circ g')^{-1} \circ g = s_{g(x)} \circ g$. Hence $g$ is an automorphism of $(M, \{s_x\})$. This completes the proof.

Ad II): Let be given $(M, \{s_x\})$ and a fixed point $o \in M$. Because $\mathrm{Aut}(M)$ is transitive on $M$ and $M$ is connected, we can see that $G$ is also transitive on $M$. Obviously, the map $\sigma$ given by (15) is an isomorphism of $G$ into $\mathrm{Aut}(M)$, and because $G$ is connected, $\sigma(G) = G$.

The isotropy group $G_o$ is closed in $G$ and $M \approx G/G_o$. Let $\pi: G \longrightarrow M$ be the canonical projection. For any $g \in G$ we have $(\pi \circ \sigma)(g) = \sigma(g)(o) = (s_o \circ g \circ s_o^{-1})(o) = s_o(g(o)) = (s_o \circ \pi)(g)$. Hence

$$\pi \circ \sigma = s_o \circ \pi \quad \text{on } G \qquad (17).$$

Let us recall that, for each automorphism $g \in G$, we have $g \circ s_x = s_{g(x)} \circ g$. In particular, for each $h \in G_o$ we obtain $\sigma(h) = s_o \circ h \circ s_o^{-1} = s_{h(o)} \circ h \circ s_o^{-1} = h$, i.e., $G_o \subset G^{\sigma}$.

Let $\underline{g}$, $\underline{g}_o$, $\underline{g}^{\sigma}$ denote the Lie algebras of $G$, $G_o$, $G^{\sigma}$ respectively. Identifying $\underline{g}$ with $T_e(G)$ we get the tangent map $\pi_*: \underline{g} \longrightarrow T_o(M)$. According to (17) we have $\pi_* \circ \sigma_* = S_o \circ \pi_*$ on $\underline{g}$ and hence

$$\pi_* \circ A = (I_o - S_o) \circ \pi_* \text{ on } \underline{g}, \quad \text{where } A = \mathrm{id} - \sigma_* \qquad (18).$$

Consider the Fitting decomposition $\underline{g} = \underline{g}_{0A} + \underline{g}_{1A}$. Clearly, $\underline{g}_o \subset \underline{g}^{\sigma} \subset \underline{g}_{0A}$. Suppose now that there is a vector $X \in \underline{g}_{0A}$, $X \notin \underline{g}_o$. Then $X' = \pi_*(X)$ is a non-zero vector of $T_o(M)$. On the other hand, $A^i(X) = 0$ for some $i$, and from (18) we obtain $(I_o - S_o)^i X' = 0$ for some $i$. Because $I_o - S_o$ is invertible (axiom (4′)), we get

$X' = 0$, a contradiction. Hence $\underline{g}_{0A} \subset \underline{g}_o$, and consequently, $\underline{g}_o = \underline{g}^{\sigma} = \underline{g}_{0A}$. Hence (i) and (ii) from Definition II.23 follow.

Finally, because $g \in G$ are automorphisms of $(M,\{s_x\})$, we get, in particular, $s_{\pi(g)} = s_{g(o)} = g \circ s_o \circ g^{-1}$, and (17) means that $s_o = s$. Thus the symmetries $\{s_x\}$ are expressed by means of (14).

Ad III): We shall start with the following

**Proposition II.26.** If $(G,H,\sigma)$ is a regular homogeneous s-manifold, then the canonical connection $\tilde{\nabla}$ of the reductive homogeneous space $G/H = M$ is invariant with respect to all transformations $s_x$ given by (14).

**Proof.** Because $G$ acts on $(M,\tilde{\nabla})$ as a group of affine transformations, it is sufficient to prove that $s$ is an affine transformation. Let us consider the Fitting decomposition $\underline{g} = \underline{m} + \underline{h}$, where $\underline{m} = \underline{g}_{1A}$, $\underline{h} = \underline{g}_{0A}$. Here $\underline{m}$ and $\underline{h}$ are invariant with respect to $A = id - \sigma_*$ and hence they are invariant with respect to the automorphism $\sigma_*$. Let $\{Q_u\}$ be the horizontal distribution of the canonical connection in $L(M)$. According to I.6, we have $Q_u = \underline{m}\cdot u$ for each $u \in \tilde{\pi}^{-1}(o)$, and $Q_{g\cdot u} = g_*(Q_u)$ for every $u \in L(M)$ and $g \in G$; here $G$ is acting on $L(M)$ in the natural way. We are going to show that $\{Q_u\}$ is s-invariant, where $\tilde{s}\colon L(M) \longrightarrow L(M)$ is induced by $\tilde{s}$. (Cf. I.21.)

First, we obtain easily $s(g\cdot x) = \sigma(g)\cdot s(x)$ for $g \in G$, $x \in M$. Hence we derive $s_*(g\cdot X) = \sigma(g)\cdot s_*(X)$ for $X \in T(M)$ and then $\tilde{s}(g\cdot u) = \sigma(g)\cdot \tilde{s}(u)$ for $u \in L(M)$. Further, we get $\tilde{s}_*(X\cdot u) = \sigma_*(X)\cdot\tilde{s}(u)$ for $X \in \underline{g}$, $u \in L(M)$. Now, for $u \in \tilde{\pi}^{-1}(o)$ we have $\tilde{s}(u) \in \tilde{\pi}^{-1}(o)$, and $\tilde{s}_*(Q_u) = \tilde{s}_*(\underline{m}\cdot u) = \sigma_*(\underline{m})\cdot\tilde{s}(u) = \underline{m}\cdot\tilde{s}(u) = Q_{\tilde{s}(u)}$. For a general $u \in L(M)$ we have $u = g\cdot u_o$ with $u_o \in \tilde{\pi}^{-1}(o)$, and $\tilde{s}_*(Q_u) = \tilde{s}_*(g_*(Q_{u_o})) =$
$\sigma(g)_*(\tilde{s}_*(Q_{u_o})) = \sigma(g)_*(Q_{\tilde{s}(u_o)}) = Q_{\sigma(g)\cdot\tilde{s}(u_o)} = Q_{\tilde{s}(g\cdot u_o)} = Q_{\tilde{s}(u)}$. □

Proof of III): $G$ acts as a group of automorphisms on the corresponding regular s-manifold $(M,\{s_x\})$ and hence $S$ is G-invariant. Consequently, $S$ is parallel with respect to $\tilde{\nabla}$. (See II.20, I.11.) According to the uniqueness part of Theorem II.17, our connection coincides with that given by (11). This completes the proof. □□

**Corollary II.27.** The canonical connection $\tilde{\nabla}$ of a connected regular s-manifold $(M,\{s_x\})$ is always complete and it satisfies $\tilde{\nabla}\tilde{R} = \tilde{\nabla}\tilde{T} = 0$. $(M,\tilde{\nabla})$ is an affine reductive space.

II.28. At the same time, we have completed the proof of Theorem II.4.

-------

We shall close this paragraph by clarifying the structure of the Lie algebra Der(M) of a connected regular s-manifold $(M,\{s_x\})$.

**Proposition II.29.** Every derivation on $(M,\{s_x\})$ is complete and thus it determines a one-parameter group of automorphisms of $(M,\{s_x\})$. Consequently, Der(M) is the Lie algebra of Aut(M).

Proof. It is well-known that every infinitesimal affine transformation with respect to a complete connection is complete ([KN I]).

**Proposition II.30.** Let us join to each vector $A \in \underline{g}$ the vector field $A^* \in \mathfrak{X}(M)$ given by $A^*(x) = A \cdot x$, $x \in M$. Then the map $A \longmapsto -A^*$ is an isomorphism of the Lie algebra $\underline{g}$ onto the Lie algebra Der(M).

Proof. First of all, G is the identity component of Aut(M) and thus $\underline{g}$ is isomorphic to Der(M). It is well-known that the correspondence $A \longmapsto -A^*$ is a homomorphism of the Lie algebra $\underline{g}$ into the Lie algebra $\mathfrak{X}(M)$. Further, the group G acts effectively on M and hence the map is an injection. Finally, each vector field $A^*$ generates the one-parameter subgroup $\{\exp tA\}$ of Aut(M) and thus $A^* \in$ Der(M). Hence the result follows. □

**Proposition II.31.** Let $G_o \subset G$ be the isotropy subgroup of G at $o \in M$, and let $\underline{g} = \underline{m} + \underline{g}_o$ be the corresponding Fitting decomposition of $\underline{g}$. Then for $X \in \underline{m}$ we have $X^* = L(X \cdot o)$, where the derivations $L(v)$, $v \in T(M)$, are defined as in (10), II.11.

Proof. According to (11) we have $\tilde{\nabla}_{X \cdot o} Y = [L(X \cdot o), Y](o)$ for each $Y \in \mathfrak{X}(M)$, and according to (1) from I.10, $\tilde{\nabla}_{X \cdot o} Y = [X^*, Y](o)$. Hence follows $[X^* - L(X \cdot o), Y](o) = 0$ for every $Y \in \mathfrak{X}(M)$. Because $X^* - L(X \cdot o) \in$ Der(M), we obtain easily that $X^* - L(X \cdot o) = 0$. □

## The group of transvections.

We shall define the elementary transvections of a regular s-manifold $(M,\{s_x\})$ as the automorphisms of the form $s_x \circ s_y^{-1}$, $x,y \in M$. Further, the group generated by all elementary transvections will be called <u>the group of transvections</u> of $(M,\{s_x\})$ and it will be denoted by $Tr(M,\{s_x\})$.

<u>Theorem II.32.</u> Let $(M,\{s_x\})$ be a connected regular s-manifold and $\tilde{\nabla}$ its canonical connection. Then the transvection group $Tr(M,\{s_x\})$ coincides with the transvection group $Tr(M)$ of the affine reductive space $(M,\tilde{\nabla})$. Further, $Tr(M)$ is a normal subgroup of $Aut(M)$.

<u>Proof.</u> As usual, $G$ will denote the identity component of the group $Aut(M)$ and $G_p$ the isotropy subgroup of $G$ at $p \in M$. Let $\underline{g} = \underline{m}_p \oplus \underline{g}_p$ be the Fitting decomposition of $\underline{g}$ with respect to $id - \sigma_{p*}$, where $\sigma_p : g \longrightarrow s_p \circ g \circ s_p^{-1}$, $g \in G$. According to Proposition II.24, $G/G_p$ is reductive with respect to this decomposition, and according to Theorem I.23, $Tr(M)$ coincides with the group $L$ generated by the ideal $\underline{\ell} = \underline{m}_p + [\underline{m}_p, \underline{m}_p]$ of $\underline{g}$. Notice that $\underline{m}_p$ is invariant with respect to $\sigma_{p*}$ and so is $\underline{\ell}$. Consequently, $L$ is invariant with respect to $\sigma_p$.

It is sufficient to show that $Tr(M,\{s_x\})$ is normal in $Aut(M)$ and that it coincides with the group $L$. For the sake of brevity, put $K = Tr(M,\{s_x\})$.

<u>Proposition II.33.</u> The group $K$ is transitive on $M$.

<u>Proof.</u> Notice that, in the proof of Proposition II.21, the map $\lambda(x) = s_x(p)$ can be also written in the form $\lambda(x) = (s_x \circ s_p^{-1})(p)$. □

<u>Lemma II.34.</u> Each element of $K$ can be joined to the identity $e \in G$ by a piecewise differentiable curve in $G$ which is contained in $K$.

<u>Proof.</u> (Cf. [Lo1].) Choose an origin $p \in M$ and let $X \in \underline{m}_p$. Then the curve $x_t = (\exp tX)(p)$ is a geodesic in $(M,\tilde{\nabla})$ according to Lemma I.25. Also, we get $x_t = Exp_p(tv)$, where $v = X \cdot p \in T_p(M)$ and $Exp_p$ denotes the exponential map at $p \in M$ with respect to $\tilde{\nabla}$. Hence we obtain easily

$$s_{Exp_p(tv)} \circ s_p^{-1} = \exp(tX) \circ s_p \circ \exp(-tX) \circ s_p^{-1} = \exp(tX) \circ \sigma_p(\exp(-tX)) =$$
$$= \exp(tX) \circ \exp(-t\, \sigma_{p*}(X)).$$

Here the map $X \longmapsto v$ is an isomorphism of $\underline{m}_p$ onto $T_p(M)$. Thus, each transformation of the form $s_y \circ s_p^{-1}$, where $y$ belongs to a normal neighborhood of $p$, can be joined to $e$ by a smooth arc in $G$ contained in $K$. Hence the result follows. □

<u>Proposition II.35.</u> $K$ is a connected normal Lie subgroup of $\mathrm{Aut}(M)$.

<u>Proof.</u> According to the theorem by Yamabe, or [KN I], Appendix 4, $K$ is a connected Lie subgroup of $G$. Further, for each automorphism $\Phi$ we have $\Phi \circ s_x \circ s_y^{-1} \circ \Phi^{-1} = s_{\Phi(x)} \circ s_{\Phi(y)}^{-1} \in K$. Hence $K$ is a normal subgroup of $\mathrm{Aut}(M)$. □

<u>Proposition II.36.</u> $K \subset L$.

<u>Proof.</u> The group $L$ is transitive on $M$. Thus, for every two points $p, x \in M$ there is $g \in L$ such that $g(p) = x$. Then $s_x \circ s_p^{-1} = g \circ s_p \circ g^{-1} \circ s_p^{-1} = g \circ [\sigma_p(g)]^{-1} \in L$. Because $K$ is generated by all transformations of the form $s_x \circ s_p^{-1}$, the result follows. □

<u>Proposition II.37.</u> $L \subset K$.

<u>Proof.</u> Let $\underline{k}$ be the Lie algebra of $K$. Then according to Proposition II.35, $\underline{k} \subset \underline{g}$ is invariant with respect to each $\sigma_{p*}$, where $\sigma_p : g \longmapsto s_p \circ g \circ s_p^{-1}$. Consider the Fitting decomposition of $\underline{k}$ and $\underline{g}$ with respect to $A = \mathrm{id} - \sigma_{p*}$: $\underline{k} = \underline{k}_{0A} + \underline{k}_{1A}$, $\underline{g} = \underline{g}_{0A} + \underline{g}_{1A}$, where $\underline{g}_{1A} = \underline{m}_p$, $\underline{g}_{0A} = \underline{g}_p$. Obviously, $\underline{k}_{0A} \subset \underline{g}_{0A}$, $\underline{k}_{1A} = \bigcap_i A^i(\underline{k}) \subset \bigcap_i A^i(\underline{g}) = \underline{g}_{1A}$. Because $K$ is transitive on $M$, the correspondence $X \longmapsto X \cdot p$, $X \in \underline{k}$, is a projection of $\underline{k}$ onto $T_p(M)$, and because $\underline{k}_{0A} \subset \underline{g}_p$, $\underline{k}_{1A}$ is mapped onto $T_p(M)$. Because $\underline{k}_{1A} \subset \underline{m}_p$ and $\underline{m}_p$ is isomorphic to $T_p(M)$, we obtain $\underline{m}_p = \underline{k}_{1A} \subset \underline{k}$ and thus $\underline{\ell} \subset \underline{k}$. Hence the result follows. □

This also completes the proof of Theorem II.32. □□

-------

In Theorem I.33 we stated a bijective correspondence between affine reductive spaces and reductive homogeneous spaces. We shall now modify our Theorem II.25 to obtain a bijective correspondence between regular s-manifolds and regular homogeneous s-manifolds.

<u>Proposition II.38.</u> Let $(M, \{s_x\})$ be a connected regular s-ma-

nifold and $o \in M$ a fixed point. Consider the Lie group automorphism $\sigma: \mathrm{Aut}(M) \longrightarrow \mathrm{Aut}(M)$ given by the formula $\sigma(g) = s_o \circ g \circ s_o^{-1}$. Then for each connected $\sigma$-invariant subgroup $K \subset \mathrm{Aut}(M)$ acting transitively on $M$ and for its isotropy subgroup $K_o$ at $o$, $(K, K_o, \sigma|_K)$ is a regular homogeneous s-manifold. The canonical connection of the reductive space $K/K_o$ coincides with that of the s-manifold $(M, \{s_x\})$.

Proof. Let $\underline{k}$ be the Lie algebra of $K$ and $\underline{k}_o$ the Lie algebra of $K_o$. Let $G$ and $G_o$ denote, as before, the identity component of $\mathrm{Aut}(M)$ and the isotropy subgroup of $G$ at $o$ respectively; let $\underline{g}, \underline{g}_o$ be the corresponding Lie algebras. Finally, denote $\sigma' = \sigma|_K$, $A = \mathrm{id} - \sigma_*$, $B = \mathrm{id}|_{\underline{k}} - \sigma'_*$. Thus, $B = A|_{\underline{k}}$.

The relation $(K^{\sigma'})^{\bullet} \subset K_o \subset K^{\sigma'}$ follows trivially from the relation $(G^{\sigma})^{\bullet} \subset G_o \subset G^{\sigma}$. Further, for the Fitting components we get easily $\underline{k}_{0B} = \underline{g}_{0A} \cap \underline{k} = \underline{g}_o \cap \underline{k} = \underline{k}_o$, $\underline{k}_{1B} = \bigcap_i B^i(\underline{k}) \subset \bigcap_i A^i(\underline{g}) = \underline{g}_{1A}$. Hence $(K, K_o, \sigma')$ is a regular homogeneous s-manifold. Further, $K/K_o \approx M$ because $K$ acts transitively on $M$, and hence $\dim(\underline{k}_{1B}) = \dim M = \dim(\underline{g}_{1A})$. Thus $\underline{k}_{1B} = \underline{g}_{1A} = \underline{m}$ in the usual denotation. From the construction of the canonical connection (Chapter I) we see easily that the canonical connections of $G/G_o$ and $K/K_o$ coincide on $M$. □

Proposition II.39. Let $(M, \{s_x\})$ be a connected regular s-manifold, $o \in M$ a fixed point, and let $\sigma: \mathrm{Aut}(M) \longrightarrow \mathrm{Aut}(M)$ be given by the formula $\sigma(g) = s_o \circ g \circ s_o^{-1}$. Then $\mathrm{Tr}(M)$ is the least $\sigma$-invariant subgroup of $\mathrm{Aut}(M)$ acting transitively on $M$.

Proof. $\mathrm{Tr}(M)$ is $\sigma$-invariant because it is normal in $\mathrm{Aut}(M)$. Further, if $K \subset \mathrm{Aut}(M)$ is $\sigma$-invariant and transitive on $M$, then $(M, \tilde{\nabla})$ is described by the reductive homogeneous space $K/K_o$ (see Proposition II.38). According to Theorem I.25, $\mathrm{Tr}(M)$ is a normal Lie subgroup of $K$. □

A regular homogeneous s-manifold $(G, H, \sigma)$ is said to be prime if $G/H$ is prime as a reductive homogeneous space (see Theorem I.33).

Theorem II.40. There is a (1-1) correspondence between the pointed, connected regular s-manifolds $(M, \{s_x\}, o)$ and the prime regular homogeneous s-manifolds $(G, H, \sigma)$. This correspondence is given by the relations $G = \mathrm{Tr}(M)$, $H = G_o$, and (15) in one direction, and by the relations $M = G/H$ and (14) in the other direction.

<u>The proof</u> follows from Theorems I.33, II.25 and from Propositions II.38, 39.

-------

As we know, the canonical connection $\tilde{\nabla}$ of a regular s-manifold $(M,\{s_x\})$ always satisfies $\tilde{\nabla}\tilde{T} = \tilde{\nabla}\tilde{R} = 0$. If $\tilde{T} = 0$, then the space $(M,\tilde{\nabla})$ is affine locally symmetric. The relation $\tilde{R} = 0$ is more interesting:

<u>Theorem II.41.</u> Let $(M,\{s_x\})$ be a regular s-manifold and $\tilde{\nabla}$ its canonical connection. If $M$ is connected and $\tilde{R} = 0$, then the transvection group $Tr(M,\{s_x\})$ is solvable and it is a covering manifold of $M$.

<u>Proof.</u> According to Theorem II.32, $Tr(M,\{s_x\})$ coincides with the transvection group $K = Tr(M)$. Let us choose a fixed $o \in M$. According to Theorem II.40 we get a regular homogeneous s-manifold $(K,K_o,\sigma)$; here $\sigma(g) = s_o \circ g \circ s_o^{-1}$ on $K$ and $K/K_o \approx M$. We have a decomposition $\underline{k} = \underline{k}_o + \underline{m}$, $\dim \underline{m} = \dim M$. Because $K/K_o$ is prime, we have $[\underline{m},\underline{m}]_{\underline{k}_o} = \underline{k}_o$. According to Formula (3) from I.13, $\tilde{R} = 0$ implies that $[[X,Y]_{\underline{k}_o},Z] = 0$ for $X,Y,Z \in \underline{m}$, and thus $[\underline{k}_o,\underline{m}] = 0$.

Now, $K$ is effective on $M$, and according to Proposition I.2, the linear isotropy representation of $K_o$ in $T_o(M)$ is faithful. Hence the adjoint representation of $\underline{k}_o$ in $\underline{m}$ is also faithful, and $[\underline{k}_o,\underline{m}] = 0$ implies $\underline{k}_o = 0$. Because $K_o$ is a discrete subgroup of $K$, we conclude that $K$ is a covering space of $M$. Finally, $\sigma$ induces an automorphism $\sigma_*$ of the Lie algebra $\underline{k} = \underline{m}$, where $\sigma_*$ has no non-zero fixed vector. According to Lemma 0.22, $\underline{k}$ is solvable, and $K$ is also solvable, because it is connected. □

<u>Corollary II.42.</u> Let $(M,\{s_x\})$ be a connected and simply connected regular s-manifold such that $\tilde{R} = 0$. Then $M$ is diffeomorphic to a solvable Lie group and homeomorphic to a Euclidean space.

<u>Proof.</u> According to Theorem II.41, $M$ is diffeomorphic to a connected and simply connected solvable Lie group. But the last one is ho-

meomorphic to a Euclidean space. (See e.g. Séminaire "Sophus Lie", 1954/55 Théorie des algèbres de Lie/Topologie des groupes de Lie, Paris 1955.)

## The s-manifolds of finite order.

Definition II.43. Let $k \geq 2$ be an integer. A regular s-manifold $(M, \{s_x\})$ is said to have order $k$ if $(s_x)^k = \text{id}$ for all $x \in M$, and $k$ is the least integer with this property. A regular homogeneous s-manifold $(G, H, \sigma)$ is said to have order $k$ if $\sigma^k = \text{id}$, and $k$ is the least integer with this property.

It is obvious that, in Theorems II.25 and II.40, the regular s-manifolds of order $k$ and the regular homogeneous s-manifolds of order $k$ correspond to each other. We are going to show that, in the case of finite order, the definitions II.2 and II.23 can be simplified.

Proposition II.44. Let $(M, \mu)$ be a manifold with a differentiable multiplication such that the maps $s_x \colon M \to M$, $x \in M$, given by $s_x(y) = x \cdot y$ satisfy the following axioms:

(1) $s_x(x) = x$,

($2^k$) $(s_x)^k = \text{identity}$, $k$ = the minimum number of this property,

(3) $s_x \circ s_y = s_z \circ s_x$, where $z = s_x(y)$,

(4) the fixed point $x$ of $s_x$ is isolated.

Then $(M, \mu)$ is a regular s-manifold of order $k$.

Proof. It is sufficient to show that (1)-(4) imply (2') and (4') of Definition II.2. (2') is a consequence of ($2^k$). Further, assuming (1) and ($2^k$) we obtain easily that (4) is equivalent to (4'). We only have to repeat the reasoning which preceded Definition II.2. □

Proposition II.45. Let $G$ be a connected Lie group, $H$ its closed subgroup and $\sigma$ an automorphism of $G$ such that

(i) $(G^\sigma)^o \subset H \subset G^\sigma$

($\text{iii}^k$) $\sigma^k = \text{identity}$ ($k$ being the minimum number with this property).

Then the triplet $(G, H, \sigma)$ is a regular homogeneous s-manifold of order $k$.

Proof. Let $\sigma_*$ be the induced automorphism of the Lie algebra $\underline{g}$ of G, and put $A = \mathrm{id} - \sigma_*$. We have to show that $\underline{g}_{0A} = \underline{h}$. Here obviously $\underline{h} = \mathrm{Ker}(A)$ and hence $\underline{h} \subset \underline{g}_{0A}$. (Cf. formula (13).) Suppose now that there is $X \in \underline{g}_{0A}$ such that $X \notin \underline{h}$. We can assume, without loss of generality, $AX \neq 0$, $A^2X = 0$. Then we get $\sigma_*(X) = X - Z$, where $\sigma_*(Z) = Z$. Hence we obtain by the induction $(\sigma_*)^2X = X - 2Z, \dots, (\sigma_*)^kX = X - kZ$. Because $(\sigma_*)^kX = X$, we get $Z = 0$, a contradiction. This completes the proof. □

Remark II.46. It is obvious that the regular s-manifolds of order 2 are nothing but the symmetric spaces in the sense of O.Loos.

Problem II.47. Can one obtain any reasonable theory of "regular s-manifolds" starting from the axioms (1),(2′),(3) and (4) ?

## ✱ Riemannian regular s-structures.

In the course of the classification of generalized symmetric Riemannian spaces of dimension $n \leq 5$ (cf. Chapter VI) the following remarkable observation was made: if $(M,g)$ is a g.s. Riemannian space which is of order $k > 2$ and irreducible, then all Riemannian regular s-structures on $(M,g)$ have the same canonical connection (which is different from the Riemannian connection). The relevant conjecture is at hand: our special observation might be a general law.

For the Riemannian symmetric spaces such a conjecture would be false: there are Riemannian symmetric spaces admitting non-parallel regular s-structures, such as $E^5$, $S^5$ and $S^2 \times E^3$ (cf. Note 4). Thus the assumption concerning the order $(k > 2)$ is essential. The irreducibility is also essential: consider a product $S^5 \times (M,g)$, where $S^5$ is the standard sphere and $(M,g)$ is a g.s. Riemannian space of order $k > 2$. Then the product is also of order $k' > 2$; yet it admits more than one canonical connection generated by its Riemannian regular s-structures.

In this paragraph we shall prove some elementary facts somehow supporting our general conjecture.

Lemma II.48. Let $\{s_x\}, \{s'_x\}$ be Riemannian regular s-structures on a connected $(M,g)$. Suppose that there exists a connected subgroup $K \subset I(M,g)$ such that
(a) $\mathrm{Tr}(M,\{s_x\}) \subset K \subset \mathrm{Aut}(M,\{s_x\})$, $\mathrm{Tr}(M,\{s'_x\}) \subset K \subset \mathrm{Aut}(M,\{s'_x\})$,
(b) $K$ is a normal subgroup of both $\mathrm{Aut}(M,\{s_x\})$ and $\mathrm{Aut}(M,\{s'_x\})$.

Then the canonical connections $\tilde{\nabla}$ and $\tilde{\nabla}'$ coincide.

Proof. Consider the automorphisms $\sigma : \mathrm{Aut}(M,\{s_x\}) \longrightarrow \mathrm{Aut}(M,\{s_x\})$, $\sigma' : \mathrm{Aut}(M,\{s'_x\}) \longrightarrow \mathrm{Aut}(M,\{s'_x\})$ given by $\sigma(g) = s_o \circ g \circ s_o^{-1}$, $\sigma'(g') = s'_o \circ g' \circ s_o'^{-1}$, where $o \in M$ is a fixed point. Then K is invariant with respect to both $\sigma$ and $\sigma'$, and according to Proposition II.38 we get regular homogeneous s-manifolds $(K,K_o,\sigma|_K)$, $(K,K_o,\sigma'|_K)$. It is sufficient to show that the corresponding reductive homogeneous spaces coincide.

Let us consider the endomorphisms $A = I - \sigma_*$, $A' = I - \sigma'_*$ of the Lie algebra $\underline{k}$. Then we have Fitting decompositions $\underline{k} = \underline{k}_{1A} + \underline{k}_{0A}$, $\underline{k} = \underline{k}_{1A'} + \underline{k}_{0A'}$, where $\underline{k}_{0A} = \underline{k}_{0A'} = \underline{k}_o$. All that we have to prove is the equality $\underline{k}_{1A} = \underline{k}_{1A'}$.

Denote $G = I(M,g)$, and let $G_o$ be the isotropy subgroup of G at a point $o \in M$. $G_o$ is compact and thus there exists an $\mathrm{ad}(G_o)$-invariant scalar product B on the Lie algebra $\underline{g}$. In particular, $s_o, s'_o \in G_o$ and thus B is invariant with respect to $\sigma_*$, $\sigma'_*$. Let us restrict the bilinear form B to the subalgebra $\underline{k} \subset \underline{g}$ and consider the orthogonal decomposition $\underline{k} = \underline{m} + \underline{k}_o$, where $\underline{m}$ is defined as the orthogonal complement of $k_o$. This decomposition is invariant with respect to both $\sigma_*$ and $\sigma'_*$. According to the Fitting lemma we get $\underline{m} = \underline{k}_{1A} = \underline{k}_{1A'}$, and this completes the proof. □

Theorem II.49. Let $\{s_x\}, \{s'_x\}$ be Riemannian regular s-structures on a connected $(M,g)$. If some of the groups $\mathrm{Tr}(M,\{s_x\})$, $\mathrm{Cl}^\bullet(\{s_x\})$, $\mathrm{Aut}^\bullet(M,\{s_x\}) \cap I(M,g)$, $\mathrm{Aut}^\bullet(M,\{s_x\})$ coincides with any of the groups $\mathrm{Tr}(M,\{s'_x\})$, $\mathrm{Cl}^\bullet(\{s'_x\})$, $\mathrm{Aut}^\bullet(M,\{s'_x\}) \cap I(M,g)$, $\mathrm{Aut}^\bullet(M,\{s'_x\})$, then the canonical connections $\tilde{\nabla}$ and $\tilde{\nabla}'$ coincide.

Proof. In all cases we obtain a group K with the properties (a) and (b) as a common value of two terms of our two series. □

Corollary II.50. Let $(M,g)$ be a generalized symmetric Riemannian space such that the identity component of the full isometry group $I(M,g)$ is the least transitive group of isometries of $(M,g)$. Then all Riemannian regular s-structures on $(M,g)$ have the same canonical connection.

Proof: obvious.

A particular case in which our Corollary holds is a space without infinitesimal rotations, i.e., admitting only finite number of isometries leaving a point fixed.

In the connection with Theorem II.32 we obtain finally

**Corollary II.51.** Let $\{s_x\}, \{s'_x\}$ be two Riemannian regular s-structures on a connected Riemannian space $(M,g)$. Then $\{s_x\}$, $\{s'_x\}$ have the same canonical connection if and only if they have the same group of transvections.

**Problem II.52.** The following conjecture is also stimulated by an observation made in the dimensions $n \leq 5$. (Cf.[K3], Ch.13): Let $(M,g)$ be a generalized symmetric Riemannian space which is of order $> 2$ and irreducible. Let $\{s_x\}$ be a Riemannian regular s-structure on $(M,g)$ and $\tilde{\nabla}$ its canonical connection. Then $I(M,g)^{o} \subset A(M,\tilde{\nabla})$.

## * Metrizable regular s-manifolds.

A regular s-manifold $(M,\{s_x\})$ is said to be metrizable if there is a Riemannian metric $g$ on $M$ which is invariant with respect to $\{s_x : x \in M\}$. Let us denote by $Cl^a(\{s_x\})$ the closure in $A(M,\tilde{\nabla})$ of the group generated by the set $\{s_x : x \in M\}$. We have the following criterion of metrizability:

**Theorem II.53.** A connected regular s-manifold $(M,\{s_x\})$ is metrizable if and only if the isotropy subgroup of $Cl^a(\{s_x\})$ at any point is compact.

**Proof.** a) Let $g$ be a Riemannian metric on $M$ which is invariant with respect to $\{s_x\}$. Then the closure $Cl(\{s_x\})$ in $I(M,g)$ (cf. Chapter 0) coincides with $Cl^a(\{s_x\})$. (We use the compact-open topology in both cases.) Hence all isotropy subgroups of $Cl^a(\{s_x\})$ are compact.

b) Let the isotropy group $G_o$ of $G = Cl^a(\{s_x\})$ be compact for a fixed point $o \in M$. Then there is a positive scalar product $g_0$ in the tangent space $M_o$ which is invariant under $G_o$. Hence we get a Riemannian metric $g$ which is invariant with respect to $G$ and hence with respect to $\{s_x\}$. □

For s-structures of finite order we get a simpler result:

**Theorem II.54.** Let $\{s_x\}$ be a regular s-structure of finite order on a connected manifold $M$ and $\tilde{\nabla}$ its canonical connection. Then $(M,\{s_x\})$ is metrizable if and only if the holonomy group of the affine reductive space $(M,\tilde{\nabla})$ is compact.

Proof. As we know from Remark I.26, the holonomy group at any point $o \in M$ is isomorphic to the isotropy subgroup at $o$ of the group $Tr(M) = Tr(M,\{s_x\})$. Suppose this subgroup to be compact. Then we can construct a Riemannian metric $g$ which is invariant with respect to $Tr(M)$. Now, for every affine transformation $\varphi: M \longrightarrow M$, the metric $\varphi^* g$ is also invariant with respect to $Tr(M)$. In fact, $Tr(M)$ is a normal subgroup of $A(M)$ and thus, for every element $h \in Tr(M)$, there is $h' \in Tr(M)$ such that $h \circ \varphi = \varphi \circ h'$. Hence $h^*(\varphi^* g) = \varphi^*(h'^* g) = \varphi^* g$. Choose a point $o \in M$ and put $\tilde{g} = g + s_o^* g + \ldots + (s_o^*)^{k-1} g$, where $k$ is the order of $\{s_x\}$. Then $\tilde{g}$ is invariant with respect to both $s_o$ and $Tr(M)$. Because each $s_x$ is of the form $s_x = h \circ s_o \circ h^{-1}$, where $h \in Tr(M)$ and $h(o) = x$, then $\tilde{g}$ is invariant with respect to $\{s_x\}$.

The converse assertion is obvious because $Tr(M,\{s_x\}) \subset I(M,g)$ for every $\{s_x\}$-invariant Riemannian metric $g$. □

## *Disconnected regular s-manifolds.

In our Definition II.2 we do not suppose the underlying manifold $M$ to be connected. Yet, our book is devoted, in fact, to the theory of connected regular s-manifolds. As concerns the disconnected regular s-manifolds, they would require a special theory which is not treated in this volume. Such a theory might be non-trivial, as the following examples show:

Example 1. Let $(M_1,\{s_x^1\})$, $(M_2,\{s_y^2\})$ be two regular s-manifolds. Let $M$ denote the disjoint sum of $M_1$ and $M_2$, $M = M_1 \vee M_2$. Then we define a new regular s-manifold $(M,\{s_z\})$ by the following rule:

$$\text{If } z \in M_1, \text{ we put } s_z|_{M_1} = s_z^1, \quad s_z|_{M_2} = \mathrm{id}.$$
$$\text{If } z \in M_2, \text{ we put } s_z|_{M_1} = \mathrm{id}, \quad s_z|_{M_2} = s_z^2.$$

Example 2. Let $(M,\{s_x\})$ be a regular s-manifold and let $A$ be an arbitrary index set. For each $\alpha \in A$ we consider a copy $(M_\alpha,\{s_{x_\alpha}^\alpha\})$ of $(M,\{s_x\})$, and put $M^\vee = \bigvee_{\alpha \in A} M$ (the disjoint sum). Then we can define a regular s-manifold $(M^{\vee},\{s^{\vee}_x\})$ by the formula

$$s^{\vee}_{x_\alpha}|M_\beta = s^{\beta}_{x_\beta} \quad \text{for any} \quad \alpha,\beta\in A,$$

where $x_\alpha, x_\beta$ denote the copies of the same element $x\in M$ in $M_\alpha, M_\beta$ respectively.

Example 3. Let $(M_i,\{s^i_{x_i}\})(i=1,2,3)$ be 3 copies of the same regular s-manifold $(M,\{s_x\})$. Put $\bar{M} = M_1 \vee M_2 \vee M_3$, and define a regular s-structure $\{\bar{s}_x\}$ on $\bar{M}$ by the following rule:

$$\bar{s}_{x_i}(u_i) = s^i_{x_i}(u_i) \quad \text{for} \quad i=1,2,3$$

$$\bar{s}_{x_i}(u_j) = s^k_{x_k}(u_k) \quad \text{for} \quad i\neq j, \quad \text{where} \quad k\in\{1,2,3\} \quad \text{and} \quad i\neq k\neq j.$$

Example 4. Let $E^2$ be a coordinate plane in the 3-dimensional Euclidean space $E^3$, and consider the regular s-manifolds $(E^2,\{r_z\})$, $(E^3,\{s_x\})$ defined as follows:

Each $r_z$ is of the form $\bar{t}_z\circ r_o\circ \bar{t}_z^{-1}$, where $\bar{t}_z$ is a translation in $E^2$ and $r_o$ is the rotation around the origin with the (positive) angle $\pi/2$ in the plane $E^2$. Each $s_x$ is of the form $t_x\circ s_o\circ t_x^{-1}$, where $t_x$ is a translation in $E^3$ and $s_o$ is the composition of two maps: one of them is the rotation of $E^3$ around the third coordinate axis with the (positive) angle $\pi/2$, and the other is the symmetry of $E^3$ with respect to the plane $E^2$. (Cf. Chapter 0, p.28.)

Let us consider the projection $p\colon E^3 \longrightarrow E^2$, and for each $x\in E^3$, let $v_x$ denote the translation of $E^2$ in the normal direction upto the hight of $x$.

Put $M = E^3 \vee E^2$ (the disjoint sum) and let us define the symmetries $\hat{s}_w$ in $M$ by the following rules:

$$\hat{s}_x(y) = s_x(y) \qquad \text{for} \quad x,y\in E^3,$$

$$\hat{s}_u(v) = r_u(v) \qquad \text{for} \quad u,v\in E^2,$$

$$\hat{s}_x(u) = r_{p(x)}(u) \qquad \text{for} \quad x\in E^3,\ u\in E^2,$$

$$\hat{s}_u(x) = v_x(r_u(p(x))) \qquad \text{for} \quad u\in E^2,\ x\in E^3.$$

Then $\{\hat{s}_w\}$ is a regular s-structure on $M$.

The proofs are left to the reader.

---

References: Primary - [F1 - 3],[GL2],[Gr],[K3],[K7],[LO],[LP1 - 2], [Lo1 - 2],[S1 - 2].

Secondary - [J],[KN I-II],[Wi].

# CHAPTER III

## LOCALLY REGULAR s-MANIFOLDS

### * Localization of the previous theory

The introductory part of this Chapter will be based on the following standard Lemma (see e.g. [Lo1]):

**Lemma III.1.** Let $(M,\nabla)$ be a manifold with a connection, $\pi: T(M) \longrightarrow M$ the projection. Then there exists a neighborhood $N_0$ of the null section $0_M$ in $T(M)$ such that the map $\pi \times \mathrm{Exp}: v \longmapsto (\pi(v), \mathrm{Exp}_{\pi(v)} v)$ is a diffeomorphism of $N_0$ onto a neighborhood $N$ of the diagonal in $M \times M$.

Let $M$ be a smooth manifold and $N \subset M \times M$ a neighborhood of the diagonal $\Delta(M \times M)$. Denote by $p_1$, $p_2$ the projections of $M \times M$ onto $M$. Then, for each $x \in M$, the set $N|_x = p_2\{p_1^{-1}(x) \cap N\}$ is a neighborhood of $x$ in $M$.

**Definition III.2.** A local regular s-structure on a manifold $M$ is a differentiable map $\mu: N \longrightarrow M$ where $N$ is a neighborhood in $M \times M$ of the diagonal $\Delta(M \times M)$ and the maps $s_x: N|_x \longrightarrow M$ defined by $s_x(y) = \mu(x,y)$ satisfy the following axioms:

(1) $s_x(x) = x$

(2) each $s_x$ is a diffeomorphism of $N|_x$ onto a neighborhood $N'_x \subset M$.

(3) There is a neighborhood $W$ of the diagonal in $M \times M \times M$ such that for $(x,y,u) \in W$ we have

$$(s_x \circ s_y)(u) = (s_z \circ s_x)(u), \quad (\text{where } z = s_x(y)).$$

(4) The tangent map $(s_{x*})_x: T_x(M) \longrightarrow T_x(M)$ has no fixed vectors except the null vector.

**III.3.** Obviously, if $\mu: N \longrightarrow M$ is a local regular s-structure on $M$ and $N' \subset N$ is a smaller neighborhood of $\Delta(M \times M)$, then the restriction of $\mu$ to $N'$ is a new local regular s-structure $\mu'$. Further, if $U \subset M$ is an open set, then $\mu_{(U)} = \mu|_{N\cap(U\times U)}$ is a local regular s-structure on $U$.

III.4. We can see easily that for $(x,y)\in N$ the definition domains of $s_x \circ s_y$ and $s_z \circ s_x$ are open neighborhoods of $y$.

Definition III.5. A manifold $M$ together with a local regular s-structure $\mu$ is called a locally regular s-manifold. The diffeomorphisms $s_x$ are called the local symmetries of $(M,\mu)$.

We shall extend the denotation $(M,\{s_x\})$ to the locally regular s-manifolds.

Definition III.6. Let $(M,\mu)$, $(M',\mu')$ be two locally regular s-manifolds. A local diffeomorphism $\Phi: U \longrightarrow U'$ ($U \subset M$, $U' \subset M'$ connected) is called a local isomorphism of $(M,\mu)$ into $(M',\mu')$ if $(\Phi \circ \mu)(\ ,\ ) = \mu'(\Phi(\ ),\Phi(\ ))$ holds on a neighborhood $Z_U$ of the diagonal in $U\times U$. In particular, a local isomorphism of $(M,\mu)$ into $(M',\mu')$ defined by a diffeomorphism $\Phi: M \longrightarrow M'$ is called a similarity of $(M,\mu)$ onto $(M',\mu')$.

III.7. Let $(M,\mu_1)$, $(M,\mu_2)$ be two locally regular s-manifolds with the same underlying manifold $M$ and such that $\mu_1$ and $\mu_2$ have the same restriction $\mu$ to a neighborhood $N$ of the diagonal $\Delta(M\times M)$. Then the identity map $i: M \longrightarrow M$ is a similarity of $(M,\mu_1)$ onto $(M,\mu_2)$.

Proposition III.8. Let $(M,\{s_x\})$ be a locally regular s-manifold. Then each local symmetry $s_x, x\in M$, restricted to a neighborhood of $x$, is a local isomorphism of $(M,\{s_x\})$ into $(M,\{s_x\})$ (called a local automorphism of $(M,\{s_x\})$).

Proof. Let $V_x \subset N|_x$ be a neighborhood of $x$ such that $\{x\}\times V_x \times V_x \subset W$ (see axiom (3) of III.2). Then for every $(y,u)\in V_x\times V_x$ we have $(s_x\circ s_y)(u) = (s_z\circ s_x)(u)$, $z = s_x(y)$. Thus, $s_x|_{V_x}$ is a local isomorphism. (Here $U = V_x$ and $Z_U = V_x\times V_x$.) $\square$

We can strenghten the previous result as follows:

Proposition III.9. For each $p\in M$ there is a neighborhood $V_p$ such that for each $x\in V_p$, $s_x$ is a local automorphism on $V_p$.

Proof. Define $V_p$ in such a way that $V_p \times V_p \times V_p \subset W$. □

The homogeneity of a regular s-manifold is replaced in the local case by a weaker property:

Proposition III.10. Let $(M, \{s_x\})$ be a connected locally regular s-manifold. Then the semi-group of local transformations generated algebraically by all local symmetries $s_x$ acts transitively on $M$. Consequently, the pseudogroup of all local automorphisms of $(M, \{s_x\})$ is transitive.

Proof. The first part can be proved by the similar procedure as Proposition II.21; the last statement is obvious from this procedure and III.9. □

Now, for a locally regular s-manifold $(M, \{s_x\})$ let $S$ again denote the tensor field defined by $S_x = (s_{x*})_x$.

Theorem III.11. There is a unique affine connection $\tilde{\nabla}$ on $(M, \{s_x\})$ (called the canonical connection) which is invariant with respect to the germs of all local symmetries $s_x$ and such that $\tilde{\nabla} S = 0$.

The proof is the same as that of Part A of Theorem II.4 with necessary localizations. For instance, we have to introduce local derivations, which correspond to local automorphisms of $(M, \{s_x\})$.

It follows from the previous Theorem that the curvature tensor $\tilde{R}$ and the torsion tensor $\tilde{T}$ of the canonical connection are invariant with respect to $\{s_x\}$ and hence with respect to $S$: $S(\tilde{R}) = \tilde{R}$, $S(\tilde{T}) = \tilde{T}$.

The propositions 0.35, 0.36 and 0.37 apply immediately to the abstract situation:

Proposition III.12. Let $\{s_x\}$ be a local regular s-structure on a manifold $M$. Then each tensor field $P$ invariant by the germs of all $s_x$ is parallel with respect to $\tilde{\nabla}$.

Proposition III.13. Let $\{s_x\}$ be a local regular s-structure on a manifold $M$. Then with respect to the canonical connection we have $\tilde{\nabla}\tilde{R} = \tilde{\nabla}\tilde{T} = 0$, $\tilde{\nabla} S = 0$.

**Theorem III.14.** A locally regular s-manifold $(M,\{s_x\})$ admits a subordinated structure of an analytic manifold. With respect to this structure, the canonical connection $\tilde{\nabla}$ and the tensor field $S$ are analytic.

**Proposition III.15.** Let $(M,\{s_x\})$, $(M',\{s'_y\})$ be two locally regular s-manifolds and $\tilde{\nabla}$, $\tilde{\nabla}'$ their canonical connections respectively. Then a local diffeomorphism $\Phi$ of $M$ into $M'$ is a local isomorphism if and only if it is a local affine map of $(M,\tilde{\nabla})$ into $(M',\tilde{\nabla}')$ such that $\Phi_*(S) = S'$ on the domain of definition of $\Phi$.

**Proof.** Let $\Phi: U \longrightarrow U'$ be a local isomorphism. Then we get $\Phi_*(S|_U) = S'|_{U'}$. Let $\nabla'$ denote the image under $\Phi$ of the connection $\tilde{\nabla}|_U$ on $U'$. Then we obtain $0 = \tilde{\nabla}S = \Phi(\tilde{\nabla}S) = \nabla'S'$, $\nabla' = \Phi(\tilde{\nabla}) = (\Phi\circ s_x)(\tilde{\nabla}) = (s'_{\Phi(x)}\circ\Phi)(\tilde{\nabla}) = s'_{\Phi(x)}(\nabla')$. According to the uniqueness part of Theorem III.11 we have $\nabla' = \tilde{\nabla}'$ on $U'$. Hence $\Phi$ is an affine map. In particular, each $s_x$ (or $s'_y$) determines a local affine transformation of $(M,\tilde{\nabla})$ (or $(M',\tilde{\nabla}')$) in a neighborhood of $x$ (or $y$, respectively) as we see from III.8.

Conversely, let $\Phi: U \longrightarrow U'$ be an affine transformation sending $S$ into $S'$. Let $p \in U$ be fixed and let $N_p$ be a connected neighborhood of $p$ having the following properties: $N_p \times N_p \subset N$, $\Phi(N_p) \times \Phi(N_p) \subset N'$, $\mu(N_p,N_p) \subset U$; $N_p \times N_p \times N_p \subset W$, $\Phi(N_p) \times \Phi(N_p) \times \Phi(N_p) \subset W'$. For each $x \in N_p$, $s_x$ is a local automorphism in $N_p$ and $s'_{\Phi(x)}$ is a local automorphism in $\Phi(N_p)$. Thus, both $s_x$ and $s'_{\Phi(x)}$ are local affine transformations in the corresponding neighborhoods. It follows that $\Phi\circ s_x$ and $s'_{\Phi(x)}\circ\Phi$ are affine maps in $N_p$. Now, the maps $\Phi\circ s_x$ and $s'_{\Phi(x)}\circ\Phi$ have the same tangent map at $x$ because $\Phi(S_x) = S'_{\Phi(x)}$. Since $N_p$ is connected, we get $\Phi\circ s_x = s'_{\Phi(x)}\circ\Phi$ on $N_p$. In other words, $(\Phi\circ s_x)(y) = (s'_{\Phi(x)}\circ\Phi)(y)$ for every $(x,y) \in N_p \times N_p$.

Constructing a suitable neighborhood $N_p$ for every $p \in U$ we conclude that $(\Phi\circ s_x)(y) = (s'_{\Phi(x)}\circ\Phi)(y)$ holds on the set $Z_U = \bigcup_{p\in U} N_p \times N_p$. Here $Z_U$ is a neighborhood of the diagonal in $U \times U$. Thus $\Phi$ is a local isomorphism. □

**Proposition III.16.** Let $(M,\tilde{\nabla})$ be an affine manifold such that $\tilde{\nabla}\tilde{R} = \tilde{\nabla}\tilde{T} = 0$ and let $S$ be a tensor field of type $(1,1)$ on $M$ such that

(a) $S$ and $I - S$ are invertible
(b) $\tilde{\nabla}S = 0$ and $S(\tilde{R}) = \tilde{R}$, $S(\tilde{T}) = \tilde{T}$.
Then there is, up to a similarity, a unique local regular s-structure $\{s_x\}$ such that $S$ is the tangent tensor field of $\{s_x\}$ and $\tilde{\nabla}$ is the canonical connection of $(M,\{s_x\})$.

Proof. Let $N_0'$ be a neighborhood of the null section $0_M$ as in Lemma III.1. We can see easily that the set $S^{-1}N_0' = \{S^{-1}_{\pi(v)}(v), v \in N_0'\}$ is also a neighborhood of $0_M$. Put $N_0 = N_0' \cap S^{-1}N_0'$, and let $N$ be the corresponding neighborhood of the diagonal in $M \times M$. For $(x,y) \in N$ define $\mu(x,y) = (\widetilde{\mathrm{Exp}}_x \circ S_x \circ \widetilde{\mathrm{Exp}}_x^{-1})(y)$. Then $\mu$ is differentiable and each $s_x$ is given by the formula $s_x = \widetilde{\mathrm{Exp}}_x \circ S_x \circ \widetilde{\mathrm{Exp}}_x^{-1}$ on $N|_x$; here $S_x = (s_{x*})_x$. Now, because $S_x(\tilde{R}_x) = \tilde{R}_x$ and $S_x(\tilde{T}_x) = \tilde{T}_x$, $s_x$ is a local affine transformation on $N|_x$ according to Appendix B7.

For a fixed $x \in M$ put $S' = s_{x*}(S)$ on $N|_x$. Because $\tilde{\nabla}S = 0$ and $s_x$ is connection-preserving, we get $\tilde{\nabla}S' = 0$. Now, $S'_x = S_x$ and a parallel extension to $N|_x$ is unique, hence $S' = S$ on $N|_x$ and $s_x$ preserves $S$.

Now, for a fixed $p \in M$, define a connected neighborhood $N_p$ such that $N_p \times N_p \subset N$, $\mu(N_p,N_p) \times \mu(N_p,N_p) \subset N$. Then for $x,y \in N_p$ we obviously have $N_p \subset N|_y$, $s_y(N_p) \subset N|_x$, $N_p \subset N|_x$, $s_x(N_p) \subset N|_{s_x(y)}$. Hence we can deduce that, for every $x, y \in N_p$, $s_x \circ s_y$ and $s_{s_x(y)} \circ s_x$ are affine transformations on $N_p$ preserving $S$. Comparing tangent maps at $y$ we can see that $s_x \circ s_y$ and $s_{s_x(y)} \circ s_x$ coincide on $N_p$. Thus formula (3) holds for every triplet $(x,y,u)$ where $x,y,u \in N_p$. Hence we find that (3) holds in a neighborhood $W$ of the diagonal in $M \times M \times M$. We have proved that $(M,\mu)$ is a locally regular s-manifold. Now, if $\mu'$ is another local regular s-structure on $M$ inducing $\tilde{\nabla}$ and $S$, then the identity map $i: M \longrightarrow M$ is a similarity of $(M,\mu)$ onto $(M,\mu')$ according to Proposition III.15. □

Definition III.17. Two locally regular s-manifolds $(M,\{s_x\})$, $(M',\{s'_y\})$ are said to be locally isomorphic if for every two points $x \in M$, $y \in M'$ there is a local isomorphism of a neighborhood of $x$ onto a neighborhood of $y$.

According to Proposition III.10, two connected locally regular s-manifolds $(M,\{s_x\})$, $(M',\{s'_y\})$ are locally isomorphic if there is at least one local isomorphism of $(M,\{s_x\})$ into $(M',\{s'_y\})$.

<u>Theorem III.18.</u> Let $(M,\{s_x\})$ be a connected locally regular s-manifold and $\pi: M' \longrightarrow M$ a covering manifold. Then there is, up to a similarity, a unique locally regular s-manifold $(M',\{s'_y\})$ which is locally isomorphic to $(M,\{s_x\})$ via the projection.

<u>Proof.</u> We construct the covering connection $\tilde{\nabla}'$ of $\tilde{\nabla}$ and the covering tensor field $S'$ of $S$ on the manifold $M'$. Then we use Proposition III.16 and Proposition III.15. □

## I n f i n i t e s i m a l   s - m a n i f o l d s .

Let $(M,\{s_x\})$ be a locally regular s-manifold, $\tilde{\nabla}$ its canonical connection, and $\tilde{R}$, $\tilde{T}$ the curvature tensor field and the torsion tensot field of $\tilde{\nabla}$ respectively. Let $o$ be a fixed point of $M$, and denote by $V = M_o$ the corresponding tangent space.

<u>Proposition III.19.</u> At the initial point $o$, the tensor fields $S$, $\tilde{R}$, $\tilde{T}$ satisfy the following algebraic conditions:

(i) Both $S_o$, $I_o - S_o$ are non-singular linear transformations of $V$.

(ii) For every $X, Y \in V$, the endomorphism $\tilde{R}_o(X,Y)$ acting as derivation on the tensor algebra $\mathcal{T}(V)$ satisfies $\tilde{R}_o(X,Y)(S_o) = \tilde{R}_o(X,Y)(\tilde{R}_o) = \tilde{R}_o(X,Y)(\tilde{T}_o) = 0$.

(iii) The tensors $\tilde{R}_o$ and $\tilde{T}_o$ are invariant by $S_o$.

(iv) $\tilde{R}_o(X,Y) = -\tilde{R}_o(Y,X)$, $\tilde{T}_o(X,Y) = -\tilde{T}_o(Y,X)$.

(v) The first Bianchi identity $\mathfrak{S}[\tilde{R}_o(X,Y)Z - \tilde{T}_o(\tilde{T}_o(X,Y),Z)] = 0$ holds.

(vi) The second Bianchi identity $\mathfrak{S}[\tilde{R}_o(\tilde{T}_o(X,Y),Z)] = 0$ holds.

<u>Proof.</u> (i) and (iii) are obvious, the relation $\tilde{R}_o(X,Y)(S_o) = 0$ follows from $\tilde{\nabla}S = 0$, and the remaining formulas are already known from I.16. □

We shall now prove that the tensors $S_o, \tilde{R}_o, \tilde{T}_o$ locally characterize the corresponding regular s-manifolds.

<u>Definition III.20.</u> <u>An infinitesimal s-manifold</u> is a collection $(V,S_o,\tilde{R}_o,\tilde{T}_o)$ where $V$ is a real finite-dimensional vector space and $S_o$, $\tilde{R}_o$, $\tilde{T}_o$ are tensors of types (1,1),(1,3),(1,2) respectively such that the conditions (i) - (vi) of III.19 are satisfied.

Two infinitesimal s-manifolds $(V_i, S_i, \widetilde{R}_i, \widetilde{T}_i)$, $i = 1,2$, are said to be isomorphic if there is a linear isomorphism $f: V_1 \longrightarrow V_2$ of vector spaces such that $f(S_1) = S_2$, $f(\widetilde{R}_1) = \widetilde{R}_2$, $f(\widetilde{T}_1) = \widetilde{T}_2$. The map $f$ is then called an isomorphism of $(V_1, S_1, \widetilde{R}_1, \widetilde{T}_1)$ onto $(V_2, S_2, \widetilde{R}_2, \widetilde{T}_2)$.

Because the pseudogroup of all local automorphisms acts transitively on a connected $(M, \{s_x\})$ and it leaves the tensor fields $S$, $\widetilde{R}$ and $\widetilde{T}$ invariant, we can see that for every two points $p, q \in M$ the collections $(M_p, S_p, \widetilde{R}_p, \widetilde{T}_p)$, $(M_q, S_q, \widetilde{R}_q, \widetilde{T}_q)$ are isomorphic infinitesimal s-manifolds. Therefore, we can introduce

Definition III.21. The infinitesimal model of a connected locally regular s-manifold $(M, \{s_x\})$ is the isomorphism class of infinitesimal s-manifolds $(M_p, S_p, \widetilde{R}_p, \widetilde{T}_p)$, $p \in M$.

Theorem III.22. Two connected locally regular s-manifolds $(M, \{s_x\})$, $(M', \{s'_y\})$ are locally isomorphic if and only if they have the same infinitesimal model.

Proof. It follows immediately from III.15 that "locally isomorphic" implies "having the same infinitesimal model". Let now $p \in M$, $p' \in M'$ be two points and let an isomorphism $f$ exist of $(M_p, S_p, \widetilde{R}_p, \widetilde{T}_p)$ onto $(M'_{p'}, S'_{p'}, \widetilde{R}'_{p'}, \widetilde{T}'_{p'})$. According to App. B7, there is an affine map $F$ of a connected neighborhood $U_p$ onto a neighborhood $U'_{p'}$ such that $F(p) = p'$, $F_{*p} = f$. Moreover, $f(S_p) = S'_{p'}$ and $\widetilde{\nabla} S = \widetilde{\nabla}' S' = 0$. Hence $F_*(S|_U) = S'|_{U'}$. According to III.15, $F$ is a local isomorphism. □

Remark. If $f: M_p \longrightarrow M'_{p'}$ is an isomorphism of infinitesimal s-manifolds as above, then it is induced by a local isomorphism $F$ of locally regular s-manifolds $(M, \{s_x\})$, $(M', \{s'_y\})$. Therefore, we shall speak sometimes briefly about an infinitesimal isomorphism $f: M_p \longrightarrow M'_{p'}$ between locally regular s-manifolds.

We shall now show the relationship between our local theory and the global concepts. In II.2 we have defined a regular s-manifold $(M, \{s_x\})$. The family $\{s_x\}$ is called a regular s-structure. Thus, a regular s-structure is also a local regular s-structure. Two regular s-manifolds $(M, \{s_x\})$, $(M', \{s'_y\})$ are said to be isomorphic if there is a diffeomorphism $\Phi: M \longrightarrow M'$ (called isomorphism) such that

$\Phi \circ s_x = s'_{\Phi(x)} \circ \Phi$ for all $x \in M$. It is easy to see that $(M, \{s_x\})$, $(M', \{s'_y\})$ are isomorphic if and only if they are similar as locally regular s-manifolds.

<u>Proposition III.23.</u> Let $(M, \{s_x\})$, $(M', \{s'_y\})$ be connected and simply connected regular s-manifolds. Then each local isomorphism of $(M, \{s_x\})$ into $(M', \{s'_y\})$ can be extended to an isomorphism.

<u>Proof.</u> Let $\Phi_U: U \longrightarrow U'$ be a local isomorphism ($U \subset M$, $U' \subset M'$ connected according to III.6). Then $\Phi_U$ is a local affine map of $(M, \tilde{\nabla})$ into $(M', \tilde{\nabla}')$ which maps $S|_U$ onto $S'|_{U'}$. According to Appendix B2, $\Phi_U$ can be extended to a global affine map $\Phi: M \longrightarrow M'$. Because $S$ is parallel with respect to $\tilde{\nabla}$ and $S'$ is parallel with respect to $\tilde{\nabla}'$, then $\Phi(S) = S'$ on $M$. Hence, the extended map $\Phi$ is a similarity, and consequently, an isomorphism. □

<u>Proposition III.24.</u> Let $(M, \{s_x\})$ be a connected and simply connected locally regular s-manifold such that its canonical connection $\tilde{\nabla}$ is complete. Then the germs of $\{s_x\}$ can be extended to the symmetries of a regular s-structure on $M$.

<u>Proof.</u> Each $s_x$, when restricted to a neighborhood of $x$, is a local affine map and the latter can be extended to a global affine transformation of $(M, \tilde{\nabla})$. This family of transformations preserves the tensor field $S$ and hence this is the wanted regular s-structure. □

<u>Convention.</u> In the following, we shall identify the infinitesimal s-manifolds, and also the regular s-manifolds, with the corresponding isomorphism classes (if there is no risk of confusion).

<u>Theorem III.25.</u> Each infinitesimal s-manifold $(V, S_o, \tilde{R}_o, \tilde{T}_o)$ is the infinitesimal model of a unique connected and simply connected regular s-manifold $(M, \{s_x\})$.

<u>Proof.</u> Using Theorem I.17 we can construct a simply connected manifold $(M, \tilde{\nabla})$ with a complete affine connection such that $\tilde{\nabla}\tilde{T} = \tilde{\nabla}\tilde{R} = 0$ and such that, for a fixed point $x \in M$, there is a linear isomorphism $f: V \longrightarrow M_x$ mapping the tensors $\tilde{T}_o$, $\tilde{R}_o$ into $\tilde{T}_x$ and $\tilde{R}_x$ respectively. Put $S_x = f(S_o)$. $(M, \tilde{\nabla})$ is an affine reductive space. According to Proposition I.32, the tensor $S_x$ can be extended to a parallel

tensor field S on M. Because $\tilde{R}_x$, $\tilde{T}_x$ are invariant under $S_x$, then $\tilde{R}$ and $\tilde{T}$ are invariant under S on M. According to III.16, M admits a locally regular s-structure $\{s_x\}$ which induces $\tilde{\nabla}$ and S. According to III.24, $\{s_x\}$ can be extended to a regular s-structure.

The uniqueness follows from III.22 and III.23. □

Corollary III.26. Each locally regular s-manifold is locally isomorphic to a regular s-manifold.

Proposition III.27. Let $(M,\{s_x\})$ be a connected locally regular s-manifold such that the canonical connection $\tilde{\nabla}$ is complete. Then $(M,\{s_x\})$ possesses a regular covering s-manifold $\pi$: $(M',\{s'_y\}) \longrightarrow (M,\{s_x\})$.

Proof. Let $M' \longrightarrow M$ be the simply connected covering manifold. Then the covering connection $\tilde{\nabla}'$ of $\tilde{\nabla}$ is complete. According to III.18, $(M,\{s_x\})$ possesses a locally regular covering s-manifold $(M',\{s'_y\})$. According to III.24, $\{s'_y\}$ can be extended to a regular s-structure on $M'$. □

In particular, each connected regular s-manifold $(M,\{s_x\})$ possesses a simply connected regular covering s-manifold $(M',\{s'_y\})$. Obviously, the deck-transformations of $(M',\{s'_y\})$ are similarities and hence automorphisms. Now we can state:

Theorem III.28. Let $(M,\{s_x\})$ be a connected locally regular s-manifold with the complete canonical connection, for example, a connected regular s-manifold. Then M is the quotient space of a simply connected regular s-manifold $M'$ factored by a group of automorphisms acting freely and properly discontinuously on $M'$. (Cf.[GL2].)

## Local regular s-triplets.

In order to establish a complete symmetry in our concepts, we shall introduce now the infinitesimal analog of a regular homogeneous s-manifold. It is a generalization of the classical notion known as "symmetric Lie algebra".

Definition III.29. By a local regular s-triplet we shall call a triplet $(\underline{g},\underline{h},\nu)$, where

(a) $\underline{g}$ is a Lie algebra, $\underline{h}$ its subalgebra and $\nu$ an automorphism of $\underline{g}$ such that $\underline{h}$ consists of all elements of $\underline{g}$ which are left fixed by $\nu$,

(b) $\underline{h} = \underline{g}_{0A}$ = the Fitting 0-component of the linear endomorphism $A = \mathrm{Id} - \nu$.

Obviously, the condition (b) is always satisfied if $\nu$ is of finite order.

It is easy to see that $(\underline{g},\underline{h},\nu)$ gives rise to a simply connected regular homogeneous s-manifold $(G,H,\sigma)$, where $\nu = \sigma_*$. In particular, if $G$ is a connected and simply connected Lie group with the Lie algebra $\underline{g}$, and $H \subset G$ is the connected subgroup corresponding to $\underline{h}$, then $H$ is the identity component of the subgroup $G^\sigma$ of all fixed elements under $\sigma$. Because $G^\sigma$ is closed, $H$ is also closed.

Having a local regular s-triplet $(\underline{g},\underline{h},\nu)$, we shall denote the Fitting 1-component $\underline{g}_{1A}$ by the usual symbol $\underline{m}$; thus we have a canonical decomposition $\underline{g} = \underline{h} + \underline{m}$, where $[\underline{h},\underline{m}] \subset \underline{m}$. (The direct proof of the last formula: obviously $\underline{h} = \mathrm{Ker}\, A$, $\underline{m} = \mathrm{Im}\, A$, and $A|_{\underline{m}}$ is an automorphism. Hence, for $X \in \underline{m}$, $U \in \underline{h}$, we have $X = AY$ for some $Y \in \underline{m}$, and $[X,U] = [Y - \nu Y,U] = [Y,U] - [\nu Y,\nu U] = A[Y,U] \in \underline{m}$.)

The triplet $(\underline{g},\underline{h},\nu)$ is said to be <u>effective</u>, if $\underline{h}$ does not contain a proper ideal of $\underline{g}$. If $(G,H,\sigma)$ is a regular homogeneous s-manifold such that $G$ acts effectively on $G/H$, then the corresponding triplet $(\underline{g},\underline{h},\sigma_*)$ is effective.

If $(\underline{g},\underline{h},\nu)$ is effective, then the map $U \longmapsto \mathrm{ad}(U)|_{\underline{m}}$ defined on $\underline{h}$ is injective. In fact, let $\underline{n}$ be the set of all elements $U \in \underline{h}$ such that $[U,\underline{m}] = 0$. For $U' \in \underline{h}$, $U \in \underline{n}$, $X \in \underline{m}$ we have $[[U',U],X] = [U',[U,X]] + [U,[X,U']] = 0$ because $[U,X] = 0$, and $[X,U'] \in \underline{m}$. For $Y \in \underline{m}$, $U \in \underline{n}$, $X \in \underline{m}$ we have trivially $[[Y,U],X] = 0$. Hence $\underline{n}$ is an ideal of $\underline{g}$, $\underline{n} \subset \underline{h}$, and consequently, $\underline{n} = (0)$.

The triplet $(\underline{g},\underline{h},\nu)$ is said to be <u>prime</u> if it is effective and satisfies $[\underline{m},\underline{m}] \cap \underline{h} = \underline{h}$. The proof of the following analog of Theorem II.40 is left to the reader:

<u>Theorem III.30.</u> There is a (1 - 1) correspondence between the infinitesimal s-manifolds $(V,S,\widetilde{R},\widetilde{T})$ and the prime local regular s-triplets $(\underline{g},\underline{h},\nu)$. This correspondence is given explicitly as follows:

(A) For a given $(V,S,\widetilde{R},\widetilde{T})$, $\underline{h}$ is the Lie subalgebra of End(V) generated by the endomorphisms $\widetilde{R}(X,Y)$, $X, Y \in V$, and $\underline{g} = V + \underline{h}$ with the multiplication given by

$$[X,Y] = [-\widetilde{T}(X,Y),\ -\widetilde{R}(X,Y)] \qquad X, Y \in V$$
$$[B,X] = BX \qquad B, C \in \underline{h},$$
$$[B,C] = BC - CB$$

finally $\nu = S \oplus id_{\underline{h}}$ and $\underline{m}$ is identified with V.

(B) For a given $(\underline{g},\underline{h},\nu)$, V is the subspace $\underline{m} = \underline{g}_{1A}$, $S = \nu|_{\underline{m}}$, $\widetilde{R}(X,Y) = -ad([X,Y]_{\underline{h}})$, $\widetilde{T}(X,Y) = -[X,Y]_{\underline{m}}$, $X, Y, Z \in \underline{m}$.

If need be, we shall always refer to this bijective correspondence as "canonical correspondence". Yet, we want to stress that, in our exposition, the concept of an infinitesimal s-manifold will be always primary and that of a local regular s-triplet only secondary.

## A construction of local regular s-structures.

We shall give now a construction of locally regular s-manifolds which is more general than that of Proposition III.16. The construction is due to A.J.Ledger and P.J.Graham, [GL2].

<u>Theorem III.31.</u> Let $(M,\nabla)$ be an affine manifold and suppose that there is a tensor field S of type (1,1) on M such that

(a) S and I - S are invertible
(b) the tensor fields R, $\nabla R$, T, $\nabla T$, $\nabla S$ and $\nabla^2 S$ are S-invariant.

Then there is, up to a similarity, a unique local regular s-structure $\{s_x\}$ on M inducing the tensor field S and such that the connection $\nabla$ is invariant by the germs of all $s_x$, $x \in M$.

<u>Proof.</u> We first deduce some lemmas.

<u>Lemma III.32.</u> Let $(M,\{s_x\})$ be a locally regular s-manifold and $\nabla$ an $s_x$-invariant affine connection on M. If a tensor field P is S-invariant and parallel with respect to $\nabla$, then it is $s_x$-invariant.

Proof. Each $s_x$ is a local affine transformation, thus it preserves the parallelism. Further, each $s_x$ preserves the tensor field $P$ at the point $x$. Because $P$ is parallel, it is invariant under the action of $s_x$. (Cf. the proof of 0.28.) □

**Lemma III.33.** Let $\nabla$ and $\tilde{\nabla}$ be affine connections on $M$ such that the tensor field $D$ defined by the relation $D_X Y = \nabla_X Y - \tilde{\nabla}_X Y$ satisfies $\tilde{\nabla} D = 0$. Then for all $X, Y \in \mathcal{X}(M)$ we have

$$\tilde{T}(X,Y) = T(X,Y) + D_Y X - D_X Y \qquad (5)$$

$$\tilde{R}(X,Y) = R(X,Y) - [D_X, D_Y] - D_{\tilde{T}(X,Y)} \qquad (6).$$

Proof. Equation (5) follows immediately from the definition of $T$ and $\tilde{T}$. Now, $\tilde{\nabla} D = 0$ means $\tilde{\nabla}_X D_Y - D_{\tilde{\nabla}_X Y} - D_Y \tilde{\nabla}_X = 0$, i.e., $[\tilde{\nabla}_X, D_Y] = D_{\tilde{\nabla}_X Y}$. Hence we get

$$\begin{aligned}
[\nabla_X, \nabla_Y] &= [\tilde{\nabla}_X + D_X, \tilde{\nabla}_Y + D_Y] = \\
&= [\tilde{\nabla}_X, \tilde{\nabla}_Y] + [\tilde{\nabla}_X, D_Y] + [D_X, \tilde{\nabla}_Y] + [D_X, D_Y] = \\
&= [\tilde{\nabla}_X, \tilde{\nabla}_Y] + D_{\tilde{\nabla}_X Y} - D_{\tilde{\nabla}_Y X} + [D_X, D_Y] = \\
&= [\tilde{\nabla}_X, \tilde{\nabla}_Y] + D_{[X,Y]} + D_{\tilde{T}(X,Y)} + [D_X, D_Y].
\end{aligned} \qquad (7)$$

Since

$$\nabla_{[X,Y]} = \tilde{\nabla}_{[X,Y]} + D_{[X,Y]} \qquad (8)$$

we obtain (6) subtracting (8) from (7). □

**Lemma III.34.** Let $(M, \nabla)$ be an affine manifold and suppose that $S$ is a tensor field of type $(1,1)$ such that $S$ and $I - S$ are invertible. Let $\tilde{\nabla}$ be the affine connection defined by

$$\tilde{\nabla}_X Y = \nabla_X Y - D_X Y, \qquad D_X Y = (\nabla_{(I-S)^{-1}X} S)(S^{-1}Y) \qquad (9).$$

If $P$ and $\nabla P$ are $S$-invariant tensor fields, then $\tilde{\nabla} P = 0$.

Proof. Let $P$ be a tensor field of type $(p,q)$ and suppose $P$ and $\nabla P$ are $S$-invariant. Thus for linear differential forms $\omega^1, \dots, \omega^p$ and vector fields $X, X_1, \dots, X_q$ we have

$$P(S^*\omega^1, \dots, S^*\omega^p, X_1, \dots, X_q) = P(\omega^1, \dots, \omega^p, SX_1, \dots, SX_q) \qquad (10)$$

$$(\nabla_X P)(S^*\omega^1,\dots,S^*\omega^p,X_1,\dots,X_q) = (\nabla_{SX}P)(\omega^1,\dots,\omega^p,SX_1,\dots,SX_q) \qquad (11).$$

Here $S^*$ denotes the transpose of $S$.

By the covariant differentiation we get from (10)

$$\begin{aligned}
&(\nabla_X P)(S^*\omega^1,\dots,S^*\omega^p, X_1,\dots,X_q) + \\
&+ \sum_{r=1}^{p} P(S^*\omega^1,\dots,(\nabla_X S^*)\,\omega^r,\dots,S^*\omega^p, X_1,\dots,X_q) = \\
&= (\nabla_X P)(\omega^1,\dots,\omega^p, SX_1,\dots,SX_q) + \\
&+ \sum_{r=1}^{q} P(\omega^1,\dots,\omega^p, SX_1,\dots,(\nabla_X S)X_r,\dots,SX_q).
\end{aligned} \qquad (12)$$

Define a tensor field $(\nabla_X S)S^{-1}$ of type (1,1) by the formula $((\nabla_X S)S^{-1})Y = (\nabla_X S)(S^{-1}Y)$ for $Y \in \mathfrak{X}(M)$. For the transposed operator $S^*$ we get easily $((\nabla_X S)S^{-1})^* = S^{*-1}(\nabla_X S^*)$. Now, using our tensor field as a derivation on the algebra of all tensor fields on $M$, we get from (10), (11) and (12)

$$\begin{aligned}
&(\nabla_{(I-S)X}P)(\omega^1,\dots,\omega^p, SX_1,\dots,SX_q) = \\
&= (((\nabla_X S)S^{-1})P)(\omega^1,\dots,\omega^p, SX_1,\dots,SX_q).
\end{aligned} \qquad (13)$$

Because $S$ and $I - S$ are invertible, (13) is equivalent to $\nabla_X P = ((\nabla_{(I-S)^{-1}X} S)S^{-1})P$. In other words, we get $\nabla_X P = D_X P$, i.e., $\tilde{\nabla} P = 0$. □

<u>Proof of Theorem III.31.</u> The local symmetries $s_x$ have to be local affine transformations satisfying the initial conditions given by the tangent maps $S_x$. Hence the uniqueness part follows.

For the existence part, let $\tilde{\nabla}$ be the connection given by (9). Because $\nabla S$ and $\nabla^2 S$ are S-invariant, we obtain easily that $D$ and $\nabla D$ are S-invariant, too. According to Lemma III.34, we have $\tilde{\nabla} S = = \tilde{\nabla} D = \tilde{\nabla} R = \tilde{\nabla} T = 0$. In particular, the conditions of Lemma III.33 are satisfied and from (5) and (6) we obtain $\tilde{\nabla}\tilde{T} = 0$, $\tilde{\nabla}\tilde{R} = 0$. Also, $\tilde{R}$ and $\tilde{T}$ are S-invariant. According to III.16, there is (up to a simila rity) a unique locally regular s-manifold $(M,\{s_x\})$ having $\tilde{\nabla}$ for the canonical connection and $S$ for the tangent tensor field. Now, the tensor field $D$ is S-invariant and parallel with respect to $\tilde{\nabla}$; according to III.32 it is $s_x$-invariant. $\tilde{\nabla}$ is also $s_x$-invariant and hence $\nabla$ is $s_x$-invariant. □ □

## The Riemannian case.

Let $(M,g)$ be a Riemannian manifold. A regular s-structure on $(M,g)$ is defined as a regular s-structure on $M$ for which all symmetries are isometries. Similarly, we define a local regular s-structure on $(M,g)$. The reader can check easily that the new definitions are equivalent to those of 0.8 and 0.30 respectively.

A triplet $(M,g,\{s_x\})$, where $\{s_x\}$ is a (local) regular s-structure on $(M,g)$, is called a Riemannian (locally) regular s-manifold.

Let $(M,g,\{s_x\})$, $(M',g',\{s'_y\})$ be two Riemannian regular s-manifolds. An isomorphism of $(M,g,\{s_x\})$ onto $(M',g',\{s'_y\})$ is a diffeomorphism $\Phi: M \longrightarrow M'$ such that (i) $\Phi$ is an isometry, (ii) $\Phi$ is an isomorphism of regular s-manifolds. Similarly, we define a local isomorphism in the locally regular case.

Propositions III.8, III.9, III.10 are easily modified to the Riemannian case. The properties of the canonical connection corresponding to III.11 - III.14 have been stated already in Chapter 0. Proposition III.15 takes on the following form:

Proposition III.35. A diffeomorphism $\Phi: M \longrightarrow M'$ is an (local) isomorphism of Riemannian (locally) regular s-manifolds if and only if it is a (local) isometry sending the tensor field $S$ onto the tensor field $S'$.

The proof is analogous to that of III.15 - we only replace the canonical connection by the Riemannian metric.

Proposition III.36. Let $(M,\tilde{\nabla})$ and $S$ be as in Proposition III.16, and let $g$ be a Riemannian metric on $M$ such that $\tilde{\nabla}g = 0$, $S(g) = g$. Then the tensor field $S$ determines a unique local regular s-structure $\{s_x\}$ on $(M,g)$ (up to a similarity).

Proof. Uniqueness is obvious. Existence part: According to Proposition III.16 there is a locally regular s-manifold $(M,\{s_x\})$ corresponding to $\tilde{\nabla}$ and $S$. According to III.32, $g$ is $s_x$-invariant and hence $\{s_x\}$ is Riemannian. □

Let $(M,g,\{s_x\})$ be a connected Riemannian locally regular s-manifold, and $o \in M$ a fixed point. Then $g_o$ is invariant by $S_o$ and, for every $X, Y \in M_o$, we have $\widetilde{R}_o(X,Y)(g_0) = 0$. Hence we can introduce the following definition:

Definition III.37. A Riemannian infinitesimal s-manifold is a collection $(V, S_o, g_0, \widetilde{R}_o, \widetilde{T}_o)$ where $V$ is a vector space, $S_o$, $\widetilde{R}_o$, $\widetilde{T}_o$ are tensors of types (1,1), (1,3), (1,2) respectively such that the conditions (i)-(vi) of III.19 are satisfied, and $g_0$ is a scalar product on $V$ such that $S_o(g_0) = g_0$ and $\widetilde{R}_o(X,Y)(g_0) = 0$ for every $X, Y \in V$.

The infinitesimal model of a Riemannian locally regular s-manifold is defined analogously to the abstract case. It is easy to prove that the infinitesimal model defines a connected Riemannian locally regular s-manifold uniquely up to a local isomorphism. We can also get a different algebraic characterization:

Theorem III.38. Two connected Riemannian regular s-manifolds $(M, g, \{s_x\})$, $(M', g', \{s'_y\})$ are locally isomorphic if and only if, for at least two points $p \in M$, $p' \in M'$, there is a linear isomorphism $f\colon M_p \longrightarrow M'_{p'}$ such that $f(g_p) = g'_{p'}$, $f(S_p) = S'_{p'}$, $f(\nabla S)_p = (\nabla' S')_{p'}$, $f(R_p) = R'_{p'}$. (Here, $\nabla$, $\nabla'$, $R$, $R'$ denote the Riemannian connections and the Riemannian curvature tensor fields respectively.)

Proof. The part "only if" is trivial. As for the converse, we obtain from our initial conditions successively $f(D_p) = D'_{p'}$, $f(\widetilde{T}_p) = \widetilde{T}'_{p'}$, $f(\widetilde{R}_p) = \widetilde{R}'_{p'}$ (using formulas (9),(5) and (6)). Thus the infinitesimal models are the same and the result follows. □

Similarly to III.23, we can prove that a local isomorphism of connected and simply connected Riemannian regular s-manifolds can be extended to an isomorphism. The analog of III.24 also holds. Further, we have the following:

Theorem III.39. Each Riemannian infinitesimal s-manifold $(V, g_0, S_o, \widetilde{R}_o, \widetilde{T}_o)$ is the infinitesimal model of a unique connected and simply connected Riemannian regular s-manifold $(M, g, \{s_x\})$.

Proof. According to the proof of Theorem III.25, we can construct a connected and simply connected regular s-manifold $(M, \{s_x\})$ corresponding to the infinitesimal s-manifold $(V, S_o, \widetilde{R}_o, \widetilde{T}_o)$. Let $f\colon V \longrightarrow M_x$

be a linear isomorphism mapping $S_o, \tilde{T}_o, \tilde{R}_o$ into $S_x, \tilde{T}_x, \tilde{R}_x$ respectively. Put $g_x = f(g_0)$. According to Proposition I.32, the tensor $g_x$ can be extended to a parallel tensor field $g$ on M. Because $S_o(g_0) = g_0$, $\tilde{\nabla} g = 0$, $\tilde{\nabla} S = 0$, the metric $g$ is invariant with respect to S on the whole of M. According to III.32, $g$ is invariant with respect to all $s_x$, $x \in M$. This completes the proof. □

Now we shall prove the Riemannian version of Theorem III.31.

<u>Theorem III.40.</u> Let $(M,g)$ be a Riemannian manifold and suppose that there is a tensor field S of type (1,1) on M such that
(a) $S - I$ is invertible,
(b) the tensor fields $g$, $R$, $\nabla R$, $\nabla S$ and $\nabla^2 S$ are S-invariant.
Then there is a local regular s-structure $\{s_x\}$ on $(M,g)$ inducing S, which is unique up to a similarity.

<u>Proof.</u> Because S preserves the Riemannian metric $g$, it is invertible. Then we can apply Theorem III.31 to the Riemannian connection $\nabla$; we obtain a local regular s-structure $\{s_x\}$ on M preserving $\nabla$. Now, $g$ is S-invariant and parallel with respect to $\nabla$. According to III.32, $g$ is $s_x$-invariant, which completes the proof. □

Let us remark that III.26, III.27 and III.28 are also true for the Riemannian case.

Finally, we shall often use in calculations the following:

<u>Proposition III.41.</u> Let $(M,g,\{s_x\})$ be a Riemannian locally regular s-manifold, and $D = \nabla - \tilde{\nabla}$ the difference tensor between the Riemannian and the canonical connection. Then we have

$$2g(D_Y X, Z) = g(\tilde{T}(X,Y),Z) + g(\tilde{T}(X,Z),Y) + g(\tilde{T}(Y,Z),X) \qquad (14).$$

<u>Proof.</u> We have $\nabla g = \tilde{\nabla} g = 0$, and thus $(D_X g)(Y,Z) = 0$ for all X, Y, Z. On the other hand, we can find easily that $D_Y X - D_X Y = \tilde{T}(X,Y)$. Now it is easy to derive (14). □

<u>Corollary III.42.</u> If $\tilde{T} = 0$, then the Riemannian manifold $(M,g)$ is locally symmetric.

<u>Proof.</u> From (14) we get $D = 0$, and hence $\nabla = \tilde{\nabla}$ satisfies $\nabla R = \tilde{\nabla}\tilde{R} = 0$. □

## I n v a r i a n t   a l m o s t   c o m p l e x   s t r u c t u r e s.

We shall start with a general result for tensor fields:

Proposition III.43. Let $(M,\{s_x\})$ be a connected locally regular s-manifold, and $(V,S,\tilde{R},\tilde{T})$ its infinitesimal model. There is a (1-1) correspondence between the tensor fields on $M$ invariant by the local automorphisms of $(M,\{s_x\})$ and the tensors on $V$ invariant by the automorphisms of $(V,S,\tilde{R},\tilde{T})$.

Proof. Choose a fixed $o \in M$ and let us identify $V$ with $M_o$. For any invariant tensor field $F$ on $M$, take its evaluation $F_o$ on $M_o$. Then $F_o$ is obviously invariant by the automorphisms of $(M_o,S_o,\tilde{R}_o,\tilde{T}_o)$. (See Remark after III.22.)

Conversely, let $F_o$ be an invariant tensor on $M_o$. For each $x \in M$, let $f_x\colon M_o \longrightarrow M_x$ be a fixed infinitesimal isomorphism, and put $F_x = f_x(F_o)$. We get a tensor field $F$ which is independent of the choice of $f_x$, $x \in M$. Now, $F$ is parallel with respect to $\tilde{\nabla}$. Indeed, the parallel transport along any curve $\widehat{ox}$ in $M$ induces an infinitesimal isomorphism $f_x\colon M_o \longrightarrow M_x$ and hence $F_x$ is the parallel translate of $F_o$. We conclude that each infinitesimal isomorphism $f_{x,y}\colon M_x \longrightarrow M_y$ preserves $F$, and $\tilde{\nabla}F = 0$. By the same procedure as in the proof of III.32 we can see that $F$ is invariant on $M$. □

Definition III.44. A <u>tensor field</u> on a locally regular s-manifold $(M,\{s_x\})$ is said to be <u>invariant</u> if it is invariant with respect to all local automorphisms.

According to Proposition III.43, all invariant almost complex structures on $(M,\{s_x\})$ can be constructed from the invariant complex structures on the vector space $V$. We shall now give sufficient conditions for the existence of such structures.

Proposition III.45. Let $(V,S,\tilde{R},\tilde{T})$ be the infinitesimal model of $(M,\{s_x\})$, where $\dim M = 2m$. Let $K \subset GL(V)$ be the centralizer of $S$. Then each complex structure $J$ on $V$ belonging to the center of $K$ determines an invariant almost complex structure on $(M,\{s_x\})$.

Proof. Obviously $K \supseteq \mathrm{Aut}(V,S,\tilde{R},\tilde{T})$ and $J$ is $K$-invariant.

Proposition III.46. Let $(V,S,\tilde{R},\tilde{T})$ be the infinitesimal model of $(M,\{s_x\})$, where $\dim M = 2m$. Suppose that (i) the transformation $S$ is semi-simple, (ii) all eigenvalues of $S$ are imaginary. Then there is at least one invariant almost complex structure on $(M,\{s_x\})$.

Proof. Let $\theta_i$, $\bar{\theta}_i$, $i = 1,\dots,r$ be mutually distinct eigenvalues of $S$. Then the complexification $V^c = V \otimes C$ can be decomposed into a direct sum of complex eigenspaces: $V^c = \sum_{i=1}^{r} (W_i \oplus \bar{W}_i)$. Because $S$ is semi-simple, all elements of any eigenspace are eigenvectors of $S$. Now, for $Z \in (W_1 \oplus \dots \oplus W_r)$ we put $J(Z) = \sqrt{-1}\cdot Z$, and $J(\bar{Z}) = -\sqrt{-1}\cdot\bar{Z}$. Then $J$ is a complex structure on $V^c$ which is an extension of a complex structure on $V$. Further, if $A \in K$ is extended to a transformation of $V^c$, then $A\circ S = S\circ A$ means that $A(W_i) \subseteq W_i$, $A(\bar{W}_i) \subseteq \bar{W}_i$ for all $i = 1,\dots r$. Hence $A\circ J = J\circ A$, and the result follows from Proposition III.45. □

Remark III.47. Let $\theta_i = \rho_i(\cos t_i + \sqrt{-1}\cdot\sin t_i)$ for $i = 1,\dots,r$. Then the complex structure $J$ from the proof of III.46 is given, on each real eigenspace $V_i = V \cap (W_i \oplus \bar{W}_i)$, by the formula

$$J|_{V_i} = [S - (\rho_i \cos t_i)I/\rho_i \sin t_i, \qquad i = 1,\dots,r.$$

Obviously, our construction depends on what we denote by $\theta_i$ and $\bar{\theta}_i$, $i = 1,\dots,r$. ($2^r$ possibilities.)

Proposition III.46 has the following consequence:

Theorem III.48. Let $(M,\{s_x\})$ be a connected locally regular s-manifold such that $\{s_x\}$ is of odd order: $(s_x)^{2k+1} = \mathrm{id}$ for $x \in M$. Then an invariant almost complex structure exists on $(M,\{s_x\})$.

Proof. Because $S^{2k+1} = I$, $S$ preserves a positive scalar product on $V$ and hence it is semi-simple. Further, the eigenvalues of $S$ are complex units which are different from 1 and -1, and hence imaginary. Also, $\dim V = 2m$. □

Let us now consider a Riemannian locally regular s-manifold $(M,g,\{s_x\})$ with the infinitesimal model $(V,g,S,\tilde{R},\tilde{T})$. Then $S$ is semi-simple on $V$ and $g$ determines a Hermitian product $h$ on the complexification $V^c$. Suppose that all eigenvalues of $S$ are imaginary (i.e., complex units different from 1 and -1).

The decomposition of $V^c$ into complex eigenspaces of $S$ is orthogonal with respect to $h$. Hence the complex structure $J$ constructed on $V^c$ by the method of the proof of III.46 leaves the Hermitian product $h$ invariant. Now, $J$ gives rise to an invariant almost Hermitian structure on $M$. (Here the invariance is meant with respect to all local Riemannian automorphisms of $(M,g,\{s_x\})$.) We summarize:

Theorem III.49. Let $(M,g,\{s_x\})$ be a connected Riemannian locally regular s-manifold of even dimension, and such that $S + I$ is a non-singular transformation. Then there exists at least one invariant almost Hermitian structure on $(M,g,\{s_x\})$.

In connection with the almost complex structures we usually put the question about the integrability. In our situation we have the following simple criterion:

Proposition III.50. Let $(M,\{s_x\})$ be a connected locally regular s-manifold, and $(V,S,\tilde{R},\tilde{T})$ its infinitesimal model, $\dim V = 2m$. Let $J$ be a complex structure on $V$ invariant with respect to all automorphisms of $(V,S,\tilde{R},\tilde{T})$. Then $J$ induces an invariant complex structure on $(M,\{s_x\})$ if and only if

$$\tilde{T}(JX,JY) - J\tilde{T}(X,JY) - J\tilde{T}(JX,Y) - \tilde{T}(X,Y) = 0 \qquad (15)$$

for every $X, Y \in V$.

Proof. We shall denote the invariant almost complex structure on $M$ by the same symbol $J$. Let us recall the Nijenhuis tensor

$$N(X,Y) = [JX,JY] - J[X,JY] - J[JX,Y] - [X,Y].$$

Obviously, $[X,Y] = \tilde{\nabla}_X Y - \tilde{\nabla}_Y X - \tilde{T}(X,Y)$ for every two vector fields $X, Y$ on $M$. Applying this formula to all terms of $N(X,Y)$ and making use of the relation $\tilde{\nabla}J = 0$ we see easily that (15) is equivalent to $N(X,Y) = 0$. □

Corollary III.51. If a locally regular s-manifold $(M,\{s_x\})$ satisfies $\widetilde{T} = 0$ then each invariant almost complex structure on $(M,\{s_x\})$ is integrable.

In the Riemannian case we are interested in the existence of Kählerian structures.

Proposition III.52. Let $(M,g,\{s_x\})$ be a connected Riemannian locally regular s-manifold, and $(V,g,S,\widetilde{R},\widetilde{T})$ its infinitesimal model; let $\dim V = 2m$. Let $J$ be a complex structure on $V$ preserving the metric and invariant with respect to all infinitesimal automorphisms. Then $J$ induces an invariant Kählerian structure on $(M,g,\{s_x\})$ if and only if $DJ = 0$, where $D$ is the tensor defined on $V$ by formula (14), III.41.

Proof. As usual, having an invariant tensor on $V$, we shall denote by the same symbol the corresponding $\widetilde{\nabla}$-parallel tensor field on $M$. Thus, the tensors $D$ and $J$ define the difference tensor field $\nabla - \widetilde{\nabla}$ and an almost Hermitian structure $J$ on $M$ respectively. The relation $DJ = 0$ on $V$ extends to the manifold $M$ by $\widetilde{\nabla}$-parallelism. Consequently, $DJ = 0$ holds on $V$ if and only if $\nabla J = 0$ holds on $M$. Thus $DJ = 0 \Longleftrightarrow (M,g,J)$ is Kählerian. $\square$

Example III.53. Consider the Riemannian regular s-manifolds of order 3 and of dimension 4 from Chapter VI (see Theorem VI.3 and its proof). Each of these spaces is defined by an infinitesimal s-manifold $(V,g,S,\widetilde{R},\widetilde{T})$, where $V$ is a 4-dimensional vector space and $g,S,\widetilde{R},\widetilde{T}$ are given, up to an isomorphism, in the following way: $S$ admits an orthogonal basis of complex eigenvectors, $\{U_1,U_2,\bar{U}_1,\bar{U}_2\}$, such that $\|U_1\| = 1$, $\|U_2\| = 1/\wp$, $\widetilde{T}(U_1,U_2) = \bar{U}_1$, $\widetilde{T}(U_j,\bar{U}_k) = 0$, $\widetilde{R}(U_2,\bar{U}_2)U_1 = U_1$, $\widetilde{R}(U_2,\bar{U}_2)\bar{U}_1 = -\bar{U}_1$, $\widetilde{R}(U_2,\bar{U}_2)U_2 = -2U_2$, $\widetilde{R}(U_2,\bar{U}_2)\bar{U}_2 = 2\bar{U}_2$, and $\widetilde{R}(Z,Z') = 0$ for other pairs of eigenvectors.

Here $V^c = (U_1,U_2) \oplus (\bar{U}_1,\bar{U}_2)$ is the decomposition into eigenspaces of $S$ $(S^2 + S + I = 0)$. Let us first define a complex structure $J$ on $V^c$ in the "standard" way described in III.46: $J(U_i) = \sqrt{-1}\cdot U_i$, $J(\bar{U}_i) = -\sqrt{-1}\cdot\bar{U}_i$. Then (15) shows that the corresponding invariant almost Hermitian structure is not integrable.

On the other hand, $\widetilde{T}$ maps $V^c \times V^c$ onto the subspace $(U_1,\bar{U}_1)$ of $V^c$. Hence $(U_1,\bar{U}_1)$, and also the orthogonal complement $(U_2,\bar{U}_2)$, is an invariant subspace of $V^c$ with respect to all infinitesimal automorphisms. Hence a new invariant complex structure $J'$ can be defined on $V^c$ by the relations $J'(U_1) = \sqrt{-1}\cdot U_1$, $J'(U_2) = -\sqrt{-1}\cdot U_2$, $J'(\bar{U}_1) = -\sqrt{-1}\cdot\bar{U}_1$, $J'(\bar{U}_2) = \sqrt{-1}\cdot\bar{U}_2$. We see easily from (14) that the subspaces $(U_1,\bar{U}_2)$, $(\bar{U}_1,U_2)$ are invariant with respect to the derivation $D$ and hence $DJ' = 0$. Hence $J'$ gives rise to an invariant Kählerian structure.

Analogously to Corollary III.51 we obtain from III.52 and (14):

Corollary III.54. If a Riemannian locally regular s-manifold $(M,g,\{s_x\})$ satisfies $\tilde{T} = 0$, then each invariant almost Hermitian structure on $(M,g,\{s_x\})$ is Kählerian.

Example III.53 shows that some invariant almost Hermitian structures can be Kählerian even if $\tilde{T} \neq 0$.

Remark III.55. We can also introduce weakly invariant tensor fields as those which are invariant under all local symmetries but not necessarily under all local automorphisms. For instance, the Hermitian structure of a Hermitian symmetric space is weakly invariant but not invariant in the sense of Definition III.44. (Cf.[KN II].)

If $(M,\{s_x\})$ is a connected and simply connected regular s-manifold, and $(V,S,\tilde{R},\tilde{T})$ its infinitesimal model, then weakly invariant tensor fields on $M$ correspond to tensors on $V$ which are invariant under $S$ and under all endomorphisms $\tilde{R}(X,Y)$, $X, Y \in V$.

Remark III.56. Ledger and Pettit, [LP2], have classified all Riemannian regular s-manifolds $(M,g,\{s_x\})$ with integrable tensor field $S$. Here a tensor field $S$ of type $(1,1)$ is called integrable if the generalized Nijenhuis tensor

$$N_S(X,Y) = S^2[X,Y] - S[SX,Y] - S[X,SY] + [SX,SY]$$

vanishes. Let $(M,\{s_x\})$ be a locally regular s-manifold. By a similar procedure as in the proof of Proposition III.50 we can prove that $N_S(X,Y) = 0$ holds if and only if $S(I-S)\tilde{T}(X,Y) = 0$. (Here we can use the additional property $S(\tilde{T}) = \tilde{T}$ which is not generally satisfied by a complex structure $J$.) Thus, $S$ is integrable if and only if $\tilde{T} = 0$. We shall return to this topic in the next Chapter.

---

References: [F1-3],[GL2],[K1],[K3],[KL],[LO],[LP2],[Lo1],[WoG].

# CHAPTER IV

## OPERATIONS WITH s-MANIFOLDS

### * Submanifolds and foliations.

Let $(M,\nabla)$ be a manifold with an affine connection. A submanifold $N \subset M$ is said to be <u>autoparallel</u> if for every curve $\gamma$: $\langle 0,1\rangle \longrightarrow N$ and every tangent vector $u \in N_{\gamma(0)}$ the parallel displacement of $u$ along $\gamma$ yields a tangent vector to $N$, i.e., $h_\gamma u \in N_{\gamma(1)}$ (cf. [KN II]).

If $N \subset M$ is autoparallel, then the connection $\nabla$ on $M$ naturally induces a connection $\nabla^N$ on $N$ as follows: for $p \in N$, $u \in T_p(N)$, $Y \in \mathfrak{X}(N)$ we put $\nabla^N_u Y = \nabla_u \tilde{Y}$, where $\tilde{Y}$ is any extension of $Y$ from a neighborhood of $p$ in $N$ to a neighborhood of $p$ in $M$.

<u>Definition IV.1.</u> Let $(M,\{s_x\})$ be a regular s-manifold. A submanifold $N \subset M$ is said to be <u>invariant</u> if the following holds: for every two points $p, q \in N$ and every local automorphism $\varphi$ of $M$ such that $\varphi(p) = q$ we have $\varphi(N \cap \varphi^{-1}(M)) \subset N$.

<u>Proposition IV.2.</u> Every invariant submanifold $N \subset M$ is autoparallel with respect to the canonical connection $\tilde{\nabla}$ of $(M,\{s_x\})$.

<u>Proof</u>. Let $N \subset M$ be invariant under all local automorphisms. Let $\gamma$: $\langle 0,1\rangle \longrightarrow N$ be a curve and $h_\gamma$: $M_{\gamma(0)} \longrightarrow M_{\gamma(1)}$ the parallel displacement along $\gamma$. Then there is a transvection $g \in Tr(M,\{s_x\})$ such that $g(\gamma(0)) = \gamma(1)$ and $g_* = h_\gamma$ on $M_{\gamma(0)}$. (Cf. Theorem II.32.) Now, because $N$ is invariant, we get $g_*(N_{\gamma(0)}) = N_{\gamma(1)}$ and hence $h_\gamma N_{\gamma(0)} = N_{\gamma(1)}$. □

<u>Proposition IV.3.</u> Let $N \subset M$ be an invariant submanifold of $(M,\{s_x\})$. Then $N$ is naturally a regular s-manifold $(N,\{s_y|_N\})$ and its canonical connection coincides with the induced connection $\tilde{\nabla}^N$.

<u>Proof</u>. Obviously, $\tilde{\nabla}^N$ is invariant under all $s_y|_N$, $y \in N$, and $\tilde{\nabla}^N(S|_N) = 0$. Now, we use the uniqueness part of Theorem II.4, (A). □

$(N,\{s_y|_N\})$ is called <u>an invariant submanifold of</u> $(M,\{s_x\})$.

Let $(V,S,\tilde{R},\tilde{T})$ be an infinitesimal s-manifold. An infinitesimal s-manifold $(W,S',\tilde{R}',\tilde{T}')$ is called <u>a submanifold of</u> $(V,S,\tilde{R},\tilde{T})$ if $W \subset V$, $S' = S|_W$, $\tilde{R}' = \tilde{R}|_W$, $\tilde{T}' = \tilde{T}|_W$. Let us remark that each submanifold of $(V,S,\tilde{R},\tilde{T})$ is uniquely defined by a subspace $W \subset V$ such that $S(W) \subset W$, $\tilde{R}(W,W)W \subset W$, $\tilde{T}(W,W) \subset W$. Further, a subspace $W \subset V$ (or, a submanifold $(W,S',\tilde{R}',\tilde{T}')$ of $(V,S,\tilde{R},\tilde{T})$ respectively) is said to be <u>invariant</u> if it is invariant with respect to all automorphisms of $(V,S,\tilde{R},\tilde{T})$.

<u>Proposition IV.4</u>. Let $(M,\{s_x\})$ be a connected regular s-manifold and $(V,S,\tilde{R},\tilde{T})$ its infinitesimal model. If $(N,\{s_y|_N\})$ is an invariant submanifold of $(M,\{s_x\})$ then its infinitesimal model is canonically an invariant submanifold $(W,S',\tilde{R}',\tilde{T}')$ of $(V,S,\tilde{R},\tilde{T})$.

<u>Proof</u>. First, let us choose $p \in N \subset M$ and put $V = T_p(M)$, $W = T_p(N)$. From Proposition IV.3 we obtain that $(W,S_p',\tilde{R}_p',\tilde{T}_p')$ is a submanifold of $(V,S_p,\tilde{R}_p,\tilde{T}_p)$, and the invariance of $W \subset V$ follows from the invariance of $N \subset M$ at the point $p$. Our formulation in terms of the infinitesimal models is now justified by the invariance of $W \subset V$. □

<u>Theorem IV.5</u>. Let $(M,\{s_x\})$ be a connected regular s-manifold and $(V,S,\tilde{R},\tilde{T})$ its infinitesimal model. Let $(W,S',\tilde{R}',\tilde{T}')$ be an m-dimensional invariant submanifold of $(V,S,\tilde{R},\tilde{T})$. Then there is an m-dimensional involutive distribution $\Delta$ on $M$ the maximal integral manifolds of which are invariant submanifolds of $(M,\{s_x\})$ with the infinitesimal model $(W,S',\tilde{R}',\tilde{T}')$.

<u>Proof</u>. Let us define the distribution $\Delta$ as follows: for each $p \in M$ choose an isomorphism $f\colon (V,S,\tilde{R},\tilde{T}) \longrightarrow (M_p,S_p,\tilde{R}_p,\tilde{T}_p)$ and put $\Delta_p = f(W)$. Then $\Delta$ is invariant with respect to all local automorphisms of $(M,\{s_x\})$. Now, the parallel displacement along $\gamma = \widehat{pq}$ with respect to $\tilde{\nabla}$ is induced by a transvection $\varphi$ such that $\varphi(p) = q$. Because every transvection is an automorphism, $\varphi_*$ sends $\Delta_p$ into $\Delta_q$, and hence $\Delta_q$ is the effect of the parallel transport of $\Delta_p$ along $\gamma$. We conclude that the distribution $\Delta$ is parallel with respect to the connection $\tilde{\nabla}$. Let now $X$, $Y$ be two local vector fields on $M$ belonging to the distribution $\Delta$. Then $[X,Y] = \tilde{\nabla}_X Y - \tilde{\nabla}_Y X - \tilde{T}(X,Y)$.

Here $\tilde{\nabla}_X Y$ and $\tilde{\nabla}_Y X$ belong to $\Delta$ because $\Delta$ is parallel, and $\tilde{T}(X,Y)$ belongs to $\Delta$ because $W$ is a submanifold of $(V,S,\tilde{R},\tilde{T})$. Thus $[X,Y]$ belongs to $\Delta$, and $\Delta$ is involutive. The maximal integral submanifolds are clearly invariant. □

Remark IV.6. The set of maximal integral submanifolds of $\Delta$ will be called the foliation corresponding to $(W,S',\tilde{R}',\tilde{T}')$.

-------

We shall now consider the Riemannian case. If we understand the notions "automorphism", "regular s-manifold", "infinitesimal s-manifold" etc. in the Riemannian sense, then the exact analogies to Propositions IV.3, IV.4 and Theorem IV.5 hold. IV.5 can be completed as follows:

Corollary IV.7. Let $(M,g,\{s_x\})$ be a connected Riemannian regular s-manifold, and $(V,g,S,\tilde{R},\tilde{T})$ its infinitesimal model. Then each m-dimensional invariant submanifold $(W,g',S',\tilde{R}',\tilde{T}')$ of $(V,g,S,\tilde{R},\tilde{T})$ determines a foliation of $(M,g)$ into m-dimensional submanifolds, which are generalized symmetric Riemannian spaces.

In general, the leaves may or may not be totally geodesic submanifolds of $(M,g)$. The following criterion is true:

Proposition IV.8. Let $\mathcal{F}$ be a foliation of $(M,g)$ into m-dimensional leaves as in Corollary IV.7. Then the foliation $\mathcal{F}$ is totally geodesic if and only if for every $X \in W$, $Z \in W^{\perp}$ we have $g(\tilde{T}(X,Z),X) = 0$.

Proof. Let $D = \nabla - \tilde{\nabla}$ denote the difference tensor between the Riemannian and the canonical connection on $M$. Then a leaf $N \in \mathcal{F}$ is totally geodesic in $(M,g)$ if, and only if, for each two vector fields $X,Y \in \mathfrak{X}(N)$ we have $\nabla_X Y \in \mathfrak{X}(N)$. Because $N$ is always autoparallel with respect to the canonical connection $\tilde{\nabla}$, we have $\tilde{\nabla}_X Y \in \mathfrak{X}(N)$. Hence $N$ is totally geodesic if and only if $D_X Y \in \mathfrak{X}(N)$ for every $X,Y \in \mathfrak{X}(N)$. Identifying $W$ with a tangent space $N_p$, we get hence the condition $D_X Y \in W$ for $X,Y \in W$. Conversely, if $D_X Y \in W$ for $X,Y \in W$, then for the invariant distribution $\Delta$ on $M$ determined by

W we get: if X,Y belong to $\Delta$, then $D_X Y$ belongs to $\Delta$, too. Hence we see that all leaves are totally geodesic.

Now, the previous algebraic condition can be reduced to the following one: $D_X X \in W$ for every $X \in W$. Indeed, for $X, Y \in W$ we get $D_X Y + D_Y X = D_{X+Y}(X+Y) - D_X X - D_Y Y \in W$, and $D_X Y - D_Y X = \widetilde{T}(X,Y) \in W$, because W is a submanifold of $(V,g,S,\widetilde{R},\widetilde{T})$. Finally, from (14), III.41, we obtain $g(D_X X, Z) = g(\widetilde{T}(X,Z), X)$ identically. For $X \in W$, $Z \in W^{\perp}$ our condition follows. □

<u>Example IV.9a</u>: Consider the 3-dimensional example from Chapter 0: $M = R^3[a,b,c]$, $ds^2 = e^{2c}da^2 + e^{-2c}db^2 + \lambda^2 dc^2$, and the typical symmetry at o is given by $a' = -b$, $b' = a$, $c' = -c$. The corresponding infinitesimal s-manifold is $(V,g,A,\widetilde{R},\widetilde{T})$, where V has an orthogonal basis $\{X,Y,Z\}$ such that $\|X\| = \|Y\| = 1$, $\|Z\| = \lambda$; further, $\widetilde{R} = 0$, $\widetilde{T}(X,Y) = 0$, $\widetilde{T}(X,Z) = -X$, $\widetilde{T}(Y,Z) = Y$, and A is given by $AX = -Y$, $AY = X$, $AZ = -Z$. Now, $W = (X,Y)$ is an invariant submanifold. The corresponding 2-dimensional foliation of M is formed by the surfaces c = const; the leaves are euclidean planes. The foliation is not totally geodesic.

<u>Example IV.9b</u>: Consider the 4-dimensional example III.53. $(U_1,\bar{U}_1)$, $(U_2,\bar{U}_2)$ are invariant submanifolds of $(V^c,g,S,\widetilde{R},\widetilde{T})$. Hence we obtain two complementary foliations $\mathcal{F}_1$ and $\mathcal{F}_2$. Representing our space as $R^4[x,y,u,v]$ with the metric given in VI.3, then $\mathcal{F}_1$ consists of the surfaces x = const., y = const., and $\mathcal{F}_2$ consists of the surfaces u = const, v = const. $\mathcal{F}_1$ is a foliation into euclidean planes and $\mathcal{F}_2$ is a foliation into hyperbolic planes. Finally, $\mathcal{F}_2$ is totally geodesic and $\mathcal{F}_1$ is not.

<u>IV.10</u>. Analogously to III.55, we can define the weak invariance as follows: a submanifold $N \subset (M,\{s_x\})$ is said to be <u>weakly invariant</u> if it is invariant with respect to all symmetries $s_x$. Let $(V,S,\widetilde{R},\widetilde{T})$ be the infinitesimal model of $(M,\{s_x\})$, and $(G,H,\sigma)$ the corresponding prime regular s-manifold. Then a subspace $W \subset V$ is said to be <u>weakly invariant with respect to</u> $(M,\{s_x\})$ if it is invariant with respect to S and to the group ad(H) acting on V. Finally, W is said to be <u>weakly invariant</u>, if it is invariant with respect to S and to all endomorphisms $\widetilde{R}(X,Y)$, $X,Y \in V$, of V. Then all previous results of this section remain valid if we replace the notion

"invariant" by the notion "weakly invariant with respect to $(M,\{s_x\})$" or, for a simply connected $(M,\{s_x\})$, by the notion "weakly invariant".

Caution: an involutive distribution $\Delta$ on $(M,\{s_x\})$ constructed for a weakly invariant $W \subset V$ is not unique in general. Each of such distributions can be obtained as follows: we take a fixed point $p \in M$, an isomorphism $f: (V,S,\tilde{R},\tilde{T}) \longrightarrow (M_p,S_p,\tilde{R}_p,\tilde{T}_p)$ and we put $\Delta_p = f(W)$. Then $\Delta_p$ determines a unique $\tilde{\nabla}$-parallel distribution $\Delta$. The reader can prove easily that $\Delta$ is weakly invariant.

Example IV.11: Let us consider a 5-dimensional g.s. space of order 4 given as $R^5[u,v,x,y,z]$ with the metric $g = du^2 + dv^2 + dx^2 + dy^2 + \varphi^2(udx + vdy - dz)^2$, $\varphi > 0$, (cf. VI, Type 1). A typical symmetry at the origin is given by the relations $u' = -v$, $v' = u$, $x' = y$, $y' = -x$, $z' = -z$. The corresponding infinitesimal model $(V,g,S,\tilde{R},\tilde{T})$ can be given as follows: $V$ is spanned onto an orthogonal basis $\{X_1,Y_1,X_2,Y_2,W\}$, where $\|X_j\| = \|Y_j\| = 1$, $\|W\| = \varphi$, $\tilde{T}(X_1,X_2) = -\tilde{T}(Y_1,Y_2) = W$, $\tilde{T}(Z,Z') = 0$ for other basic vectors, $\tilde{R} = 0$; $SX_1 = -Y_1$, $SY_1 = X_1$, $SX_2 = Y_2$, $SY_2 = -X_2$, $SW = -W$.

Now, the subspaces $(X_1,Y_1)$, $(X_2,Y_2)$, $(W)$ are all weakly invariant. The corresponding foliations $\mathcal{F}_1$, $\mathcal{F}_2$, $\mathcal{F}_3$ are mutually orthogonal and given as follows:

$$\mathcal{F}_1: x,y,z = \text{const}, \quad \mathcal{F}_2: u,v,z = \text{const}, \quad \mathcal{F}_3: x,y,u,v = \text{const}.$$

The leaves of $\mathcal{F}_1$ and $\mathcal{F}_2$ are totally geodesic euclidean planes, the foliation $\mathcal{F}_3$ consists of geodesic lines. Nevertheless, the space $(R^5,g)$ is irreducible.

## Decomposition of s-manifolds.

Definition IV.12. A regular s-manifold $(M,\{s_x\})$ is called the direct product of regular s-manifolds $(M_1,\{s^1_u\})$, $(M_2,\{s^2_v\})$ if $M$ is diffeomorphic to $M_1 \times M_2$, and (under an identification $M = M_1 \times M_2$ via a diffeomorphism)

$$s_{(u,v)}(p,q) = (s^1_u(p),\ s^2_v(q)) \quad \text{for every } (u,v),(p,q) \in M.$$

$(M,\{s_x\})$ is called reducible or irreducible according to whether it is a direct product or not.

Proposition IV.13. Let $(M,\{s_x\}) = (M_1,\{s_u^1\}) \times (M_2,\{s_v^2\})$, and denote by $\tilde{\nabla}$, $\tilde{\nabla}_1$, $\tilde{\nabla}_2$ the canonical connections on $M$, $M_1$, $M_2$ respectively. Then $(M,\tilde{\nabla}) = (M_1,\tilde{\nabla}_1) \times (M_2,\tilde{\nabla}_2)$ (direct product of affinely connected spaces). Further, $\mathrm{Tr}(M,\{s_x\}) = \mathrm{Tr}(M_1,\{s_u^1\}) \times \times \mathrm{Tr}(M_2,\{s_v^2\})$ (direct product of Lie groups).

Proof. We can show by the same way as in the proof of IV.3 that the product connection $\tilde{\nabla}_1 \times \tilde{\nabla}_2$ on $M$ is the canonical one. The last assertion is obvious from the definitions. □

Definition IV.14. An infinitesimal s-manifold $(V,S,\tilde{R},\tilde{T})$ is called the direct sum of infinitesimal s-manifolds $(V_i,S_i,\tilde{R}_i,\tilde{T}_i)$, $i=1,2$, if $V = V_1 + V_2$ (direct sum) and $S(X) = \sum_{i=1}^{2} S_i(\pi_i X)$, $\tilde{R}(X,Y)Z = \sum_{i=1}^{2} \tilde{R}_i(\pi_i X,\pi_i Y)\pi_i Z$, $\tilde{T}(X,Y) = \sum_{i=1}^{2} T_i(\pi_i X,\pi_i Y)$, where $\pi_i\colon V \longrightarrow V_i$ are projections.

An infinitesimal s-manifold is called reducible or irreducible according to whether it is a direct sum or not.

IV.15. Let $(M,\{s_x\}) = (M_1,\{s_u^1\}) \times (M_2,\{s_v^2\})$. Then each submanifold $M'_{1p} = M_1 \times \{p\}$ or $M'_{2,q} = \{q\} \times M_2$ of $(M,\{s_x\})$ is weakly invariant but not necessarily invariant (cf. IV.10). We see from Theorem IV.5 that $\{M'_{1p}\}$, $\{M'_{2q}\}$ are invariant foliations if and only if $\mathrm{Aut}(M,\{s_x\}) = \mathrm{Aut}(M_1,\{s_u^1\}) \times \mathrm{Aut}(M_2,\{s_v^2\})$. Also, if $(V,S,\tilde{R},\tilde{T})$ is a direct sum of $(V_i,S_i,\tilde{R}_i,\tilde{T}_i)$, $i=1,2$, then $V_1$ and $V_2$ are weakly invariant submanifolds of $(V,S,\tilde{R},\tilde{T})$. The proof is left to the reader.

Proposition IV.16. A connected, simply connected s-manifold $(M,\{s_x\})$ is reducible if and only if its infinitesimal model is reducible.

Proof: it follows easily from IV.13 and Theorem III.25.

IV.17. In the Riemannian case, we can give the analogous definitions and obtain the similar results. Here we only require, in addition, that the Riemannian metric also decomposes. The details are left to the reader.

As a special case of reducibility we have the following

**Proposition IV.18.** Let $(V,S,\tilde{R},\tilde{T})$ be an infinitesimal s-manifold such that $\tilde{T}=0$, and let $V = \sum_{\alpha\in A} V_\alpha$ be the decomposition of $V$ into the eigenspaces of $S$. (Here $A$ denotes the set of all real eigenvalues and all couples of mutually conjugate complex eigenvalues.) Then $(V,S,\tilde{R},\tilde{T})$ decomposes into a direct sum of infinitesimal s-manifolds $(V_\alpha, S_\alpha, \tilde{R}_\alpha, \tilde{T}_\alpha)$, $\alpha\in A$. The same result is true in the Riemannian case.

**Proof.** In the Riemannian case, the eigenspaces are always mutually orthogonal. It suffices to show that $\tilde{R} = \sum_{\alpha\in A} \tilde{R}\circ\pi_\alpha$. From the relations $S(\tilde{R}) = \tilde{R}$ and $\tilde{R}(X,Y)\cdot S = 0$ we get $\tilde{R}(SX,SY) = \tilde{R}(X,Y)$ for each $X, Y\in V$. Hence $\tilde{R}(\pi_\alpha X, \pi_\beta Y) = 0$ whenever $\alpha\neq\beta$. Further, $\tilde{R}(\pi_\alpha X,\pi_\alpha Y)\pi_\beta Z = -\tilde{R}(\pi_\beta Z,\pi_\alpha X)\pi_\alpha Y - \tilde{R}(\pi_\alpha Y,\pi_\beta Z)\pi_\alpha X = 0$ according to the first Bianchi identity (here $\tilde{T}=0$). Hence the result follows. □

As an application, we get the following (cf. [LP2])

**Theorem IV.19** (Ledger, Pettit). Let $(M,g,\{s_x\})$ be a connected and simply connected Riemannian regular s-manifold with an integrable tensor field $S$. Let $n_o$ be the multiplicity of the eigenvalue $-1$ of $S$, and let $n_1,\dots,n_r$ be the dimensions of the eigenspaces corresponding to mutually different pairs of imaginary eigenvalues. Then

$$(M,g,\{s_x\}) = (M_o,g_0,\{s_u^o\})\times(M_1,g_1,\{s_v^1\})\times\dots\times(M_r,g_r,\{s_w^r\}),$$

$$\dim M_i = n_i,\quad i = 0,1,\dots,r,$$

where $(M_o,g_0,\{s_u^o\})$ is a Riemannian symmetric space with geodesic symmetries $s_u^o$, and each $(M_i,g_i)$, $i=1,\dots,r$, is a Kählerian symmetric space.

**Proof.** We have $\tilde{T}=0$ according to III.56, and hence the above decomposition follows from IV.16 and IV.18. It suffices to prove the last assertion.

On each $(M_i,g_i,\{s^i_{u_i}\})$, $i=1,\dots,r$, we can construct an invariant almost Hermitian structure $J_i$ according to III.49. Because $\tilde{T}=0$, each $(M_i,g_i)$ is locally symmetric and Kählerian (III.42 and

III.54). Being simply connected and complete, each $(M_i, g_i)$ is Riemannian symmetric. Finally, each almost complex structure $J_i$ is parallel with respect to the Riemannian connection $\nabla_i$, and according to III.32, the Kähler structure is invariant with respect to the usual symmetries. □

Problem IV.20. If a Riemannian symmetric space $(M,g)$ is a direct product, then its family of geodesic symmetries decomposes as a direct product, too. For non-parallel regular s-structures it sometimes happens - but rather rarely - that a simply connected Riemannian regular s-manifold $(M,g,\{s_x\})$ is irreducible whereas the underlying Riemannian manifold $(M,g)$ is a direct product. For example, in Note 4 we have constructed a non-parallel and irreducible regular s-structure on $S^2 \times E^3$.

Find some general algebraic criteria !

## * Periodic s-manifolds.

Let $(M,\{s_x\})$ be a connected regular s-manifold. $(M,\{s_x\})$ will be said to be of semi-simple (or solvable) type according to whether the group $Tr(M,\{s_x\})$ is semi-simple (or solvable) respectively.

The following theorem generalizes a classical one:

Theorem IV.21. Each connected periodic (i.e. finite order) regular s-manifold $(M,\{s_x\})$ is canonically a fibre bundle such that the base space is a regular s-manifold of semi-simple type, and the fibres are invariant submanifolds, which are regular s-manifolds of solvable type.

Proof. Let $(V,S,\tilde{R},\tilde{T})$ be the infinitesimal model of $(M,\{s_x\})$, $(G,H,\sigma)$ the corresponding prime regular homogeneous s-manifold and $(\underline{g},\underline{h},\nu)$ the corresponding prime local regular s-triplet, $\nu = \sigma_*$. (Cf. Chapter III.) Here $G = Tr(M,\{s_x\})$.

Let us recall the equivariant version of the Levi theorem (see e.g. [KN II]).

Lemma IV.22. Let $K$ be a compact group of automorphisms of a Lie algebra $\underline{g}$, then $\underline{g}$ has a Levi subalgebra $\underline{s}$ which is invariant by $K$; i.e., we have a $K$-invariant vector space decomposition $\underline{g} = \underline{r} + \underline{s}$, where $\underline{r}$ is the radical of $\underline{g}$ and $\underline{s}$ is a semi-simple subalgebra.

(The decomposition $\underline{g} = \underline{r} + \underline{s}$ is called a Levi decomposition.)

Now, let $\underline{g} = \underline{r} + \underline{s}$ be a Levi decomposition which is invariant with respect to $\nu$ (recall that $\nu$ is supposed to be periodic). Consider the Fitting decomposition $\underline{g} = V + \underline{h}$ with respect to $A = \mathrm{id} - \nu$. Both $\underline{r}$ and $\underline{s}$ are invariant with respect to $A$; using their Fitting decompositions we obtain readily $\underline{r} = \underline{r} \cap V + \underline{r} \cap \underline{h}$, $\underline{s} = \underline{s} \cap V + \underline{s} \cap \underline{h}$, and $\underline{h} = \underline{r} \cap \underline{h} + \underline{s} \cap \underline{h}$. Now, $(\underline{r}, \underline{r} \cap \underline{h}, \nu|_{\underline{r}})$, $(\underline{s}, \underline{s} \cap \underline{h}, \nu|_{\underline{s}})$ are local regular s-triplets (they need not be prime nor effective). The connected subgroup $R \subset G$ determined by $\underline{r}$ is the radical of $G$; it is a normal $\sigma$-invariant subgroup, and it is also closed in $G$ (cf. [KN II]). The triplet $(R, R \cap H, \sigma|_R)$ is a regular homogeneous s-manifold, and it determines a regular s-manifold $(N, \{s'_y\})$. Obviously, the infinitesimal model of $(N, \{s'_y\})$ is $(W, S|_W, \tilde{R}|_W, \tilde{T}|_W)$, where $W = \underline{r} \cap V$; it is easy to see that the last quadruplet is an invariant submanifold of $(V, S, \tilde{R}, \tilde{T})$ and hence it determines an invariant foliation of $(M, \{s_x\})$ into submanifolds isomorphic to $(N, \{s'_y\})$. Further, $R$ is solvable and acts (possibly not effectively) as a group of automorphisms on $(N, \{s'_y\})$. Because $\mathrm{Tr}(N, \{s'_y\})$ is a normal subgroup of $\mathrm{Aut}(N, \{s'_y\})$, it is a solvable group, too.

Consider the factor group $S = G/R$. Then $S$ is semi-simple because its Lie algebra is isomorphic to $\underline{s}$. Further, the triplet $(G/R, H/(R \cap H), \tilde{\sigma})$, where $\tilde{\sigma}$ is the projection of $\sigma$ on $G/R$, is a regular homogeneous s-manifold. In fact, the corresponding local triplet $(\underline{g}/\underline{r}, \underline{h}/(\underline{r} \cap \underline{h}), \tilde{\nu})$ is naturally isomorphic to $(\underline{s}, \underline{s} \cap \underline{h}, \nu|_{\underline{s}})$, which is a local regular s-triplet. The corresponding regular s-manifold is of semi-simple type - the proof is similar as above. Now, consider $G/H = M$ as a fibre bundle over $(G/R)/[H/(R \cap H)]$; the fibres are diffeomorphic to $R/(R \cap H)$, and they coincide with our invariant leaves. □

Remark IV.23. It is worth mentioning that we make a strong use of the Levi theorem when proving that the base space is a regular s-manifold (more specifically, in the proof that the projections of the symmetries onto the base are also symmetries). In the classical result for the symmetric spaces the Levi theorem is not needed.

**Theorem IV.24.** Let $(M,\{s_x\})$ be a connected periodic regular s-manifold which is metrizable. Then $(M,\{s_x\})$ admits two complementary foliations $\mathcal{F}_1$, $\mathcal{F}_2$ such that (a) $\mathcal{F}_1$ is invariant and its leaves are regular s-manifolds of solvable type, (b) $\mathcal{F}_2$ is weakly invariant and its leaves are regular s-manifolds of semi-simple type.

Proof. Let $(V,S,\tilde{R},\tilde{T})$, $(G,H,\sigma)$, $(\underline{g},\underline{h},\nu)$ have the same meaning as before. It is sufficient to prove that there is a Levi decomposition $\underline{g} = \underline{r} + \underline{s}$ such that $\underline{s} \cap V$ is a weakly invariant subspace of $V$ with respect to $(M,\{s_x\})$. Now, $H$ is isomorphic to the holonomy group of $(M,\tilde{\nabla})$, and because $(M,\{s_x\})$ is metrizable, $H$ is compact (cf. II.53). Further $\nu$ commutes with each element $ad(h)$, $h \in H$, on $\underline{g}$, and thus the group $K$ generated by $ad(H)$ and $\nu$ is compact. Let $\underline{g} = \underline{r} + \underline{s}$ be a K-invariant Levi decomposition. Because $V$ is K-invariant, $\underline{s} \cap V$ is K-invariant, too, and by definition, it is weakly invariant. □

-------

IV.25. Consider now a periodic s-manifold of the semi-simple type. A classical theorem on symmetric spaces says that a symmetric Lie algebra $(\underline{g},\underline{h},\nu)$ with $\underline{g}$ semi-simple $(\nu^2 = id)$ can be decomposed as a direct sum of simple symmetric Lie algebras and the symmetric Lie algebras of the type $(\tilde{\underline{g}} + \tilde{\underline{g}}, \Delta\tilde{\underline{g}}, \nu)$, where $\tilde{\underline{g}}$ is simple, $\Delta\tilde{\underline{g}} = \{(X,X);\ X \in \tilde{\underline{g}}\}$, and $\nu(X,Y) = (Y,X)$ for $X,Y \in \tilde{\underline{g}}$. ([H],[KN II].)

In Chapter 0 we have introduced regular homogeneous s-manifolds of the form $(G^k, \Delta G^k, \sigma)$, where $\sigma(x_1,\dots,x_k) = (x_k,x_1,\dots,x_{k-1})$, or, what is the same, local regular s-triplets of the form $(\underline{g}^k, \Delta\underline{g}^k, \nu)$, where $\nu(X_1,\dots,X_k) = (X_2,X_1,\dots,X_{k-1})$ for $X_1,\dots,X_k \in \underline{g}$. There is a natural question whether each periodic local regular s-triplet $(\underline{g},\underline{h},\nu)$ with $\underline{g}$ semi-simple can be written as a direct sum of s-triplets $(\underline{g}_i,\underline{h}_i,\nu_i)$ with $\underline{g}_i$ simple, and "cyclic" s-triplets of the form $(\underline{g}_j^{k_j}, \Delta\underline{g}_j^{k_j}, \nu_j)$ with $\underline{g}_j$ simple. The answer is negative, as the following counter-example shows:

Consider the prime local regular s-triplet $(\underline{so}(3) + \underline{so}(3), \underline{h}, \nu)$, where $\underline{so}(3)$ is spanned on a natural basis $\{X,Y,Z\}$ or $\{X',Y',Z'\}$ respectively, $\underline{h} = (X + X')$, and $\nu(X) = X'$, $\nu(Y) = -Z'$, $\nu(Z) = Y'$, $\nu(X') = X$, $\nu(Y') = -Z$, $\nu(Z') = Y$. Because $\dim \underline{h} = 1$, our s-triplet is not isomorphic with one of the "cyclic" form. (Cf. Type 5a of the classification list of generalized symmetric Riemannian spaces of dimension 5, Chapter VI.)

The lesson from here is that the study of periodic regular s-manifolds with the semi-simple basic group cannot be reduced to the study of those with a simple basic group. This is an additional difficulty in the classification problem.

## * Amalgamations.

Amalgamation is a generalized product operation which enables constructing irreducible s-manifolds from irreducible s-manifolds. We shall discuss here the Riemannian case; the abstract case is analogous but easier.

Let $\mathcal{V} = (V, g, S, \widetilde{R}, \widetilde{T})$ be a Riemannian infinitesimal s-manifold. A vector space decomposition $V = W \dotplus A$ is said to be amalgamating if

(i) $W$, $A$ are non-trivial, weakly invariant submanifolds of $\mathcal{V}$,

(ii) $W$, $A$ are mutually orthogonal,

(iii) $W$ is an ideal in the sense that $\widetilde{T}(X,Y) \in W$ whenever $X \in W$ or $Y \in W$, and $\widetilde{R}(X,Y)Z \in W$ whenever $X \in W$, or $Y \in W$, or $Z \in W$.

Such a decomposition always exists if $\mathcal{V}$ is reducible. What is more important, it often exists also for $\mathcal{V}$ irreducible.

Let now $\mathcal{V}_i = (V_i, g_i, S_i, \widetilde{R}_i, \widetilde{T}_i)$, $i = 1,2$, be two Riemannian infinitesimal s-manifolds and $V_i = W_i \dotplus A_i$ amalgamating decompositions. Then $A_1$, $A_2$ are underlying spaces of infinitesimal s-manifolds $\mathcal{A}_1 \subset \mathcal{V}_1$, $\mathcal{A}_2 \subset \mathcal{V}_2$. Suppose that there exists an isomorphism $f\colon \mathcal{A}_1 \longrightarrow \mathcal{A}_2$. Then the subspace

$$V_1 \cup_f V_2 = (W_1 \dotplus W_2) \dotplus \{Z + f(Z) \mid Z \in A_1\} \qquad (1)$$

is a weakly invariant submanifold of the direct sum $\mathcal{V}_1 \dotplus \mathcal{V}_2$. The Riemannian infinitesimal s-manifold $\mathcal{V}_1 \cup_f \mathcal{V}_2$ resulting from here is called the amalgamation of $\mathcal{V}_1$ and $\mathcal{V}_2$ via $f$. We notice that the decomposition (1) is once more amalgamating. If $\mathcal{V} = (V, g, S, \widetilde{R}, \widetilde{T})$ admits an amalgamating decomposition $V = W \dotplus A$, then we can always take two copies of $\mathcal{V}$ and construct the object $\mathcal{V} \cup_f \mathcal{V}$, where $f$ is the identity map $id\colon A \longrightarrow A$. By iterations, or also directly, we can construct a sequence of "self-amalgamations" $\mathcal{V}$, $\mathcal{V} \cup_f \mathcal{V}$, $\mathcal{V} \cup_f \mathcal{V} \cup_f \mathcal{V}$, ..., etc.

According to Theorem III.39, the operation of amalgamation applies to simply connected regular s-manifolds, as well.

<u>Example IV.26</u>: Each of the Riemannian s-manifolds we have introduced in the examples IV.9a,b and IV.11 admits an amalgamating decomposition. In particular, consider Example IV.9a, where the only amalgamating decomposition is $V = (X,Y) \dotplus (Z)$. Let us take two copies $\mathcal{V}_1$, $\mathcal{V}_2$ with different $\lambda$'s, say $\lambda_1, \lambda_2$. If we construct the amalgamation $\mathcal{V}_1 \cup_f \mathcal{V}_2$ for $f: Z_1/\lambda_1 \mapsto Z_2/\lambda_2$, and "integrate" this infinitesimal s-manifold, we obtain a 5-dimensional simply connected g.s. Riemannian space corresponding to Type 2 of our classification list (cf. Theorem VI.5), namely

$$M = R^5(x,y,z,w,t), \quad g = e^{-2\lambda_1 t}dx^2 + e^{2\lambda_1 t}dy^2 + e^{-2\lambda_2 t}dz^2 + e^{2\lambda_2 t}dw^2 + dt^2.$$

Here, $(M,g)$ is irreducible as a Riemannian manifold.

We shall prove now a general irreducibility theorem. A Riemannian infinitesimal s-manifold $(V,g,S,\widetilde{R},\widetilde{T})$ is said to be <u>centerless</u> if it does not contain a subspace $C \neq \emptyset$ such that $\widetilde{T}(Z,X) = 0$, $\widetilde{R}(Z,X) = 0$ for $Z \in C$, $X \in V$.

<u>Theorem IV.27</u>. Let $\mathcal{V}_1$, $\mathcal{V}_2$ be irreducible and centerless Riemannian infinitesimal s-manifolds. If an amalgamation $\mathcal{V}_1 \cup_f \mathcal{V}_2$ exists, then it is irreducible and centerless.

<u>Proof</u>. Denote $\mathcal{V}_i = (V_i, g_i, S_i, \widetilde{R}_i, \widetilde{T}_i)$, $V_i = W_i \dotplus A_i$, $f: A_1 \longrightarrow A_2$, $\mathcal{V}_1 \cup_f \mathcal{V}_2 = (V_1 \cup_f V_2, g, S, \widetilde{R}, \widetilde{T})$. Consider the projections $\pi_i$: $V_1 \dotplus V_2 \longrightarrow V_i$, $i = 1,2$. $\pi_1$ and $\pi_2$ restricted to $V_1 \cup_f V_2$ are still surjective. Suppose $\mathcal{V}_1 \cup_f \mathcal{V}_2$ to be reducible, i.e., $V_1 \cup_f V_2 = L_1 \dotplus L_2$ = the orthogonal sum of weakly invariant submanifolds of $\mathcal{V}_1 \cup_f \mathcal{V}_2$. Then $V_i = \pi_i(L_1) + \pi_i(L_2)$ for $i = 1,2$ (the usual sum of subspaces). Let us remark that

$$\widetilde{R}_i \circ \pi_i = \pi_i \circ \widetilde{R}, \quad \widetilde{T}_i \circ \pi_i = \pi_i \circ \widetilde{T}, \quad S_i \circ \pi_i = \pi_i \circ S \qquad (2).$$

Hence $\pi_i(L_1) \cap \pi_i(L_2)$ is in the centre $C_i$ of $\mathcal{V}_i$, and thus trivial. We get $V_i = \pi_i(L_1) \dotplus \pi_i(L_2)$ (the direct sum of vector spaces).

Let us suppose that, for instance, the decomposition $V_1 = \pi_1(L_1) \dotplus \pi_1(L_2)$ be non-trivial. From (2) we see that $\pi_1(L_1)$, $\pi_1(L_2)$ are weakly invariant submanifolds of $\mathcal{V}_1$. Further, because the tensors $S,\widetilde{R},\widetilde{T}$ split with respect to $V_1 \cup_f V_2 = L_1 \dotplus L_2$, the tensors $S_1,\widetilde{R}_1,\widetilde{T}_1$ split with respect to $V_1 = \pi_1(L_1) + \pi_2(L_2)$. It remains to prove that this decomposition is also orthogonal.

Let $X + H \in \pi_1(L_1)$, where $X \in W_1$, $H \in A_1$. If $U_1 \in L_1$ is such that $X + H = \pi_1(U_1)$, then $U_1 = X + H + f(H) + X'$, where $X' \in W_2$. Let $X = Z_1 + Z_2$ for some $Z_1 \in L_1$, $Z_2 \in L_2$. From $\pi_2(X) = 0$ we get $\pi_2(Z_1) = -\pi_2(Z_2) = 0$ since $\pi_2(L_1) \cap \pi_2(L_2)$ is trivial. Thus $Z_1, Z_2 \in W_1$. Now, $g(U_1 - Z_1, Z_2) = 0$, i.e., $g(Z_2 + H + f(H) + X', Z_2) = 0$. Because $Z_2 \in W_1$ is orthogonal to $H \in A_1$, $f(H) \in A_2$, $X' \in W_2$, we get hence $g(Z_2, Z_2) = 0$, i.e., $Z_2 = 0$. Thus $X = Z_1 \in L_1$. Similarly, we derive that $X' \in L_1$. Hence $H + f(H) \in L_1$.

If now $Y + G \in \pi_1(L_2)$, we obtain in the same way $Y \in L_2$, $G + f(G) \in L_2$. Hence $g(H + f(H), G + f(G)) = 0$, i.e., $2g_1(H,G) = 0$, and also $g_1(X,Y) = 0$. From here we obtain $g_1(X + H, Y + G) = 0$, which is the wanted orthogonality.

We conclude that $\pi_1(L_1) \dotplus \pi_1(L_2)$ is the direct sum of weakly invariant submanifolds of $\mathcal{V}_1$ - a contradiction to the irreducibility of $\mathcal{V}_i$.

So, the only possibility is that both decompositions are trivial, e.g., $\pi_2(L_1) = 0$, $\pi_1(L_2) = 0$. But then $L_1 \subset W_1$, $L_2 \subset W_2$ and $V_1 \cup_f V_2 = L_1 \dotplus L_2 \subset W_1 \dotplus W_2$, a contradiction to the assumption $A_1 \neq (0)$.

We conclude that $\mathcal{V}_1 \cup_f \mathcal{V}_2$ is irreducible. It is easy to show that it is also centerless. □

<u>Remark IV.28.</u> The infinitesimal s-manifolds from Examples IV.9a,b are irreducible and centerless. In Example IV.11, the infinitesimal s-manifold is irreducible but not centerless.

Let $\mathcal{V} = (V,g,S,\widetilde{R},\widetilde{T})$ be a Riemannian infinitesimal s-manifold, and $V = W \dotplus A$ an amalgamating decomposition. For each $\lambda > 0$ we shall consider a Riemannian infinitesimal s-manifold $\mathcal{V}^{\lambda} = (V,g^{\lambda},S,\widetilde{R},\widetilde{T})$ where $g^{\lambda} = g$ on $W$ and $g^{\lambda} = \lambda \cdot g$ on $A$. Now, if $\mathcal{V}_n = (V_n, g_n, S_n, \widetilde{R}_n, \widetilde{T}_n)$ denotes the <u>n-th self-amalgamation of $\mathcal{V}^{1/n}$</u> via $id_A$, then $\mathcal{V}$ is isomorphic to a weakly invariant submanifold of $\mathcal{V}_n$. In fact, let us take $n$ copies $\mathcal{V}^{(1)}, \ldots, \mathcal{V}^{(n)}$ of $\mathcal{V}^{1/n}$. Then

$$V_n = W^{(1)} \dotplus \ldots \dotplus W^{(n)} \dotplus \{Z^{(1)} + \ldots + Z^{(n)} \mid Z \in A\},$$

and $\mathcal{V}$ is isomorphic to the submanifold $W^{(1)} \dotplus \{Z^{(1)} + \ldots + Z^{(n)} \mid Z \in A\}$ of $\mathcal{V}_n$ via the correspondence $(X + Z) \longmapsto X^{(1)} + (Z^{(1)} + \ldots + Z^{(n)})$, $X \in W$, $Z \in A$.

Now, if $(M_n, g_n)$ is the simply connected g.s. Riemannian space corresponding to $\mathcal{V}_n$, then $\mathcal{V}$ generates a foliation $\mathcal{F}$ of $(M_n, g_n)$ in the sense of IV.7. Moreover, the criterion IV.8 is satisfied because $W^{(1)} + \ldots + W^{(n)}$ is an ideal of $\mathcal{V}_n$; hence $\mathcal{F}$ is totally geodesic.

Application IV.29 (Spaces of order 3). Let $(M,g)$ be a generalized symmetric Riemannian space of order 3, and $\{s_x\}$ a regular s-structure on $(M,g)$, $(s_x)^3 = \mathrm{id}$. Suppose that the infinitesimal model $\mathcal{V} = (V, g, S, \tilde{R}, \tilde{T})$ is irreducible, centerless and admits an amalgamating decomposition $V = W \dotplus A$. Then $\mathcal{V}$ gives rise to an infinite sequence $(M_i, g_i, \{s_{i,x}\})$, $\dim M_i \longrightarrow \infty$, of irreducible Riemannian s-manifolds, where the underlying spaces $(M_i, g_i)$ are of order 3.

Proof. For each $n = 1, 2, \ldots$ consider the infinitesimal s-manifold $\mathcal{V}_n$ and the space $(M_n, g_n)$ as above. Because $S^3 = \mathrm{Id}$, we have $(S_n)^3 = \mathrm{Id}$, and the space $(M_n, g_n)$ is 3-symmetric. According to Theorem IV.27, the s-manifold $(M_n, g_n, \{s_{n,x}\})$ is irreducible. Finally, $(M_n, g_n)$ cannot be 2-symmetric, otherwise the totally geodesic foliation $\mathcal{F}$ generated by $\mathcal{V}$ would consist of Riemannian symmetric spaces, and $(M,g)$ would be 2-symmetric, a contradiction. □

In particular, our procedure applies to Example IV.9b.

## * C o m p l e x i f i c a t i o n s .

Let $(M,h)$ ge a Hermitian manifold. A regular s-structure $\{s_x\}$ on M is said to be Hermitian if each $s_x$ is a holomorphic, metric-preserving transformation of $(M,h)$. The triplet $(M, h, \{s_x\})$ is then called a Hermitian regular s-manifold. If some Hermitian $\{s_x\}$ exists on $(M,h)$, then we call $(M,h)$ a generalized symmetric Hermitian space.

Let $(M, g, \{s_x\})$ be a connected and simply connected Riemannian regular s-manifold such that its canonical connection satisfies $\tilde{R} = 0$. According to Theorem II.41 and Cor. II.42, the group $\mathrm{Tr}(M)$ is simply transitive on $(M,g)$. The infinitesimal model $(V, g, S, 0, \tilde{T})$ is nothing but the Lie algebra $\mathfrak{k}$ of $\mathrm{Tr}(M)$ provided with a scalar product $g$ and an isometric automorphism S. Here $[X,Y] = -\tilde{T}(X,Y)$ for every $X, Y \in V$.

Let us consider the complexification $\mathfrak{l}^c = \mathfrak{l} \otimes C$ as a complex Lie algebra. If $\{X_1,\ldots,X_n\}$ is a basis of $\mathfrak{l}$, we can make $\mathfrak{l}^c$ a 2n-dimensional real Lie algebra with the basis $\{X_1,\ldots,X_n,\sqrt{-1}X_1,\ldots,\sqrt{-1}X_n\}$ and with the complex structure $J$ defined by $JX = \sqrt{-1}X$. (Cf. [KN II].) In the following, we shall put $Y_i = \sqrt{-1}X_i$ for $i = 1,\ldots,n$. Thus $JX_i = Y_i$, $JY_i = -X_i$ for all $i$. Further, for the Lie bracket in $\mathfrak{l}^c$ we get $J([X,Y]) = [JX,Y] = [X,JY]$ for every $X,Y \in \mathfrak{l}^c$. Also, $S$ extended to $\mathfrak{l}^c$ by the linearity satisfies $S^c \circ J = J \circ S^c$; it is an automorphism of $\mathfrak{l}^c$ preserving the complex structure. Finally, let us define a real Hermitian inner product $\tilde{g}^c$ on $\mathfrak{l}^c$ (considered as a 2n-dimensional real vector space with the complex structure $J$) putting $\tilde{g}^c(X_i,X_j) = g(X_i,X_j)$, $\tilde{g}^c(X_i,Y_j) = 0$, $\tilde{g}^c(Y_i,Y_j) = g(X_i,X_j)$ for every $i,j = 1,\ldots,n$. Then $\tilde{g}^c$ is invariant under $S^c$, too. (Here $\tilde{g}^c$ is not to be confused with $g^c$ - - the complexification of $g$, or with the complex Hermitian product defined on $\mathfrak{l}^c$ as an n-dimensional complex vector space.)

We have constructed a 2n-dimensional Riemannian infinitesimal s-manifold $(V^c,\tilde{g}^c,S^c,0,\tilde{T}^c)$ provided with an S-invariant Hermitian complex structure $J$. The corresponding simply connected Riemannian regular s-manifold $(M^c,\tilde{g}^c,\{s_z^c\})$ possesses a natural structure of a Hermitian regular s-manifold. In fact, according to Remark III.55, the complex structure $J$ on $V^c$ determines an $s_x$-invariant almost complex structure $J$ on $M^c$. Further, because $(V^c,J)$ was a complex Lie algebra, $(M^c,J)$ is a complex Lie group, and thus $J$ is integrable. (Cf. [KN II] p.130.)

We have proved the main part of the following

**Theorem IV.30.** Every connected and simply connected Riemannian regular s-manifold $(M,g,\{s_x\})$ satisfying $\tilde{R} = 0$ possesses a uniquely determined complexification $(M^c,\tilde{g}^c,\{s_z^c\})$. Here $(M^c,\tilde{g}^c,\{s_z^c\})$ is a Hermitian regular s-manifold such that $\dim_C M^c = \dim_R M = n$. Further, $(M^c,\tilde{g}^c,\{s_z^c\})$ admits an n-dimensional totally geodesic foliation the leaves of which are naturally isomorphic to $(M,g,\{s_x\})$.

Proof. It is sufficient to prove the last assertion. Obviously, $(V,g,S,0,\tilde{T})$ is a weakly invariant submanifold of $(V^c,\tilde{g}^c,S,0,\tilde{T}^c)$; thus the foliation $\mathcal{F}$ is well-defined (cf. IV.10). Now, $J(V)$ is the orthogonal complement of $V$ in $V^c$ with respect to $\tilde{g}^c$; hence for $X \in V$, $Z \in V^{\perp}$ we get $Z = JX'$, $X' \in V$, and $\tilde{g}^c(\tilde{T}^c(X,Z),X) = \tilde{g}^c(\tilde{T}^c(X,JX'),X) = \tilde{g}^c(J(\tilde{T}^c(X,X')),X) = 0$. According to Proposition IV.8, $\mathcal{F}$ is totally geodesic. □

Problems IV.31. Let $(M^c, g^c, \{s^c_z\})$ be the complexification of $(M,g,\{s_x\})$.

(1) Supposing that $(M,g)$ is euclidean, is $(M^c,g^c)$ also euclidean?

(2) Supposing that $(M,\{s_x\})$ is irreducible, and $\dim M > 2$, is $(M^c,\{s^c_z\})$ irreducible? (Cf. also Example IV.35.)

## * Spaces without infinitesimal rotations.

We shall say that a Riemannian manifold $(M,g)$ does not admit infinitesimal rotations if $\dim I(M,g) = \dim M$. In particular, if $(M,g)$ is connected and simply connected, without infinitesimal rotations, then the identity component $I^{\bullet}(M,g)$ of $I(M,g)$ is simply transitive on $(M,g)$, and the isotropy subgroup of $I(M,g)$ at every point is finite.

Proposition IV.32. Let $(M,g)$ be a g.s. Riemannian space without infinitesimal rotations. Then for every Riemannian regular s-structure $\{s_x\}$ on $(M,g)$ we have $Tr(M,\{s_x\}) = I^{\bullet}(M,g)$, and the canonical connection satisfies $\tilde{R} = 0$.

Proof. We have $Tr(M,\{s_x\}) \subset I^{\bullet}(M,g)$, and $Tr(M)$ is transitive on $(M,g)$. Hence $\dim Tr(M) \leq \dim I^{\bullet}(M,g) = \dim M \leq \dim Tr(M)$, and thus $Tr(M)$ is an open Lie subgroup of $I^{\bullet}(M,g)$. Hence the relation $Tr(M) = I^{\bullet}(M,g)$ follows. Because the isotropy subalgebra $\underline{h}$ of $Tr(M)$ at any point is trivial, we get $\tilde{R} = 0$. □

As a consequence, all admissible s-structures $\{s_x\}$ have the same canonical connection $\tilde{\nabla}$ - it is the Cartan $(-)$-connection of the group $I^{\bullet}(M,g)$. Further, the group $I^{\bullet}(M,g)$ is solvable, and if $M$ is simply connected, then it is homeomorphic to a Euclidean space. (Cf. II.41, 42, 50.)

The g.s. Riemannian spaces without infinitesimal rotations occur frequently. (All spaces with this property are of solvable type.) Let us have a look at our classification list. The spaces of order 4 and of dimension 3 have been discussed in Chapter 0. Now, in dimension $n = 5$, the following spaces of the classification do not admit infinitesimal rotations (cf. Chapter VI):

spaces of Type 2 for $\lambda_1 > \lambda_2$,
spaces of Type 4 for $\lambda + \bar{\lambda} \neq 0$, $\nu \neq 0$,
spaces of Type 7 for $\lambda \neq 0$
spaces of Type 9 (of order 6).

On the other hand, spaces of Type 1 have a solvable group $Tr(M)$ and they are homeomorphic to $R^5$; nevertheless the isotropy subgroup of $I^\bullet(M,g)$ at a point is isomorphic to the group $U(2)$. (See [K3].) Spaces of Type 4 for $\nu = 0$ are of solvable type (i.e., the group $I^\bullet(M,g)$ is solvable) but they posses infinitesimal rotations.

We shall now state our main result concerning complexifications:

<u>Theorem IV.33.</u> Let $(M,g)$ be a simply connected generalized symmetric Riemannian space without infinitesimal rotations. Then there is a uniquely determined g.s. Hermitian space $(M^c,\tilde{g}^c)$, called <u>the complexification of $(M,g)$</u>, with the following property: to every Riemannian regular s-structure $\{s_x\}$ on $(M,g)$ there is a Hermitian regular s-structure $\{s^c_z\}$ on $(M^c,\tilde{g}^c)$ such that $(M^c,\tilde{g}^c,\{s^c_z\})$ is the complexification of $(M,g,\{s_x\})$ in the sense of Theorem IV.30.

<u>Proof.</u> According to the procedure of Theorem IV.30, $(M^c,\tilde{g}^c)$ is uniquely determined by the Lie algebra $\mathfrak{l}$ of $Tr(M)$ and by the scalar product $g$. But according to Proposition IV.32, $\mathfrak{l}$ is also the Lie algebra of $I(M,g)$. Hence, $(M^c,\tilde{g}^c)$ is uniquely determined by $(M,g)$. □

<u>Corollary IV.34.</u> The complexification $(M^c,\tilde{g}^c)$ of $(M,g)$ is homeomorphic to a complex Cartesian space. It admits a totally geodesic foliation the leaves of which are isometric to $(M,g)$. If $(M,g)$ is not locally symmetric, then $(M^c,\tilde{g}^c)$ is not locally symmetric, too.

<u>Proof.</u> The complexification $\mathfrak{l}^c$ of the solvable Lie algebra $\mathfrak{l}$ is solvable, too. Hence the underlying manifold $M^c$ of the corresponding simply connected Lie group $L^c$ is homeomorphic to a $C^n$. This proves the first assertion. The second one follows from Theorem IV.30, and the third one is an immediate consequence of the second. □

Example IV.35. The manifold $R^3(a,b,c)$ with the Riemannian metric $ds^2 = e^{2c}da^2 + e^{-2c}db^2 + \lambda^2 dc^2$ (g.s. space of order 4) does not posses infinitesimal rotations. Its complexification is the manifold $C^3(u,v,w)$ with the Hermitian metric $ds^2 = e^{(w+\bar{w})}du d\bar{u} + e^{-(w+\bar{w})}dv d\bar{v} + \lambda^2 dw d\bar{w}$. A Hermitian symmetry at the origin is given by the relations $u' = -v$, $v' = u$, $w' = -w$. Our Hermitian manifold is reducible as a real Riemannian manifold of dimension 6; namely, it decomposes into the direct product of a 5-dimensional space from Example IV.26 and a real line. But as a Riemannian s-manifold of order 4, our Hermitian manifold is irreducible.

## Pseudo-duality.

Pseudo-duality is defined only for connected, simply connected regular s-manifolds of order 4. We shall limit ourselves to the (more important) Riemannian case.

Let $(M,g,\{s_x\})$ be a Riemannian regular s-manifold of order 4 which is connected and simply connected, and let $\mathcal{V} = (V,g,S,\tilde{R},\tilde{T})$ be its infinitesimal model. Now, $S$ has only the eigenvalues $\pm\sqrt{-1}$, $-1$. Consider the decomposition $V = W \dot{+} Z$, where $W$ is the eigenspace corresponding to $\pm\sqrt{-1}$ and $Z$ is the eigenspace corresponding to $-1$. Here $W \perp Z$, and both subspaces are S-invariant.

Consider the quintuplet $\mathcal{V}' = (V,g,S,\tilde{R}',\tilde{T}')$, where $\tilde{R}' = -\tilde{R}$, $\tilde{T}' = -\tilde{T}$ on $W \times W$, and $\tilde{R}' = \tilde{R}$, $\tilde{T}' = \tilde{T}$ on $(W \times Z) \cup (Z \times W) \cup (Z \times Z)$. We can check easily that $\mathcal{V}'$ is a Riemannian infinitesimal s-manifold. Here we use only the properties $S(\tilde{R}) = \tilde{R}$, $S(\tilde{T}) = \tilde{T}$ which imply $\tilde{T}(W \times W) \subset Z$, $\tilde{T}(W \times Z) \subset W$. Now $\mathcal{V}'$ (or the corresponding simply connected $(M',g',\{s'_y\})$) is said to be <u>pseudo-dual</u> to $\mathcal{V}$ (or to $(M,g,\{s_x\})$ respectively. Obviously, pseudo-duality is a kind of duality: if $\mathcal{V}'$ is pseudo-dual to $\mathcal{V}$, then $\mathcal{V}$ is pseudo-dual to $\mathcal{V}'$.

An equivalent definition is the following: Let $\underline{g} = V + \underline{h}$ be the Lie algebra for $\mathcal{V} = (V,g,S,\tilde{R},\tilde{T})$ as in III.30 ($\underline{h}$ is generated by the endomorphisms $\tilde{R}(X,Y)$, $X,Y \in V$). Hence $\underline{g} = W \dot{+} Z \dot{+} \underline{h}$. Let $\underline{g}^c$ be the complexification of $\underline{g}$, then $\underline{g}' = \sqrt{-1}W \dot{+} Z \dot{+} \underline{h}$ is a subalgebra of $\underline{g}^c$. We shall consider $\underline{g}'$ as a real Lie algebra, and define an inner product on $V' = \sqrt{-1}W \dot{+} Z$ by the rule $g'(\sqrt{-1}X,\sqrt{-1}Y) = g(X,Y)$ for $X,Y \in W$, $g'(\sqrt{-1}X,U) = 0$ for $X \in W$, $U \in Z$, and $g'(U,U') = g(U,U')$ for $U,U' \in Z$. Also, an automorphism $\nu$ of $\underline{g}'$ is defined by means of the automorphism $S + id_{\underline{h}}$ of $\underline{g}$. Hence we get a regular homogeneous s-manifold $(G',H,\sigma)$ of order 4 (cf.

III.29) with a left invariant metric, and the corresponding Riemannian regular s-manifold $(G'/H, g', \{s'_y\})$ is pseudo-dual to $(M, g, \{s_x\})$. The latter construction is analogous to duality in ordinary symmetric spaces; hence the notion "pseudo-duality".

If $(M, g, \{s_x\})$, $(M', g', \{s'_y\})$ are pseudo-dual one to another, then the g.s. Riemannian spaces $(M, g)$, $(M', g')$ are said to be pseudo--dual, too.

Example IV.36. In the classification list of 5-dimensional g.s. Riemannian spaces (Chapter VI), the types (5a,5b), (6a,6b) and (8a,8b) are pseudo-dual pairs (of the form elliptic - hyperbolic). The other types are self-dual when represented by corresponding Riemannian regular s-manifolds.

## P r o p e r t i e s   o f   e i g e n v a l u e s .

We shall start with the following lemma, which is well-known:

Lemma IV.37. Let $\underline{g}$ be a Lie algebra and $A: \underline{g} \longrightarrow \underline{g}$ an endomorphism. Let $\underline{g}^c = \sum_{\alpha \in I} \underline{g}_\alpha$ be the decomposition of the complexification $\underline{g}^c$ into the complex eigenspaces corresponding to different eigenvalues $\alpha$ of $A$. Here $\underline{g}_\alpha = \{Z \in \underline{g}^c \mid (A - \alpha I)^k Z = 0$ for some $k\}$.

Then
$$[\underline{g}_\alpha, \underline{g}_\beta] \subset \underline{g}_{\alpha\beta} \quad \text{if } \alpha\beta \text{ is an eigenvalue,}$$
$$[\underline{g}_\alpha, \underline{g}_\beta] = 0 \quad \text{otherwise.}$$

As an application to infinitesimal s-manifolds, we have

Proposition IV.38. Let $(V, S, \tilde{R}, \tilde{T})$ be an infinitesimal s-manifold and $V^c = \sum_{\alpha \in J} V_\alpha$ the decomposition of $V^c = V \otimes C$ into (maximal) complex eigenspaces corresponding to mutually different eigenvalues $\alpha$ of $S$. Then

(a) $\tilde{T}(V_\alpha, V_\beta) = 0$ if $\alpha, \beta \in J, \alpha\beta \notin J$

(b) $\tilde{T}(V_\alpha, V_\beta) \subset V_{\alpha\beta}$ if $\alpha, \beta, \alpha\beta \in J$

(c) $\tilde{R}(V_\alpha, V_\beta) = 0$ if $\alpha, \beta \in J, \alpha\beta \neq 1$

(d) $\tilde{R}(Z, Z')(V_\alpha) \subset V_\alpha$ for every $Z, Z' \in V^c$ and $\alpha \in J$.

Proof. Let $\underline{g} = V + \underline{h}$ be the Lie algebra for $(V, S, \tilde{R}, \tilde{T})$ as in III.30, $\underline{h}$ = the Lie algebra generated by all endomorphisms $\tilde{R}(X, Y)$, $X, Y \in V$.

Thus $[X,Y] = (-\tilde{T}(X,Y),\ -\tilde{R}(X,Y))$ $\quad X,Y \in V$

$[A,X] = AX$ $\quad A \in \underline{h},\ X \in V$

$[A,B] = AB - BA$ $\quad A,B \in \underline{h}$

Then $S^+ = S + id_{\underline{h}}$ is an automorphism of $\underline{g}$, as follows from the relations $S(\tilde{T}) = \tilde{T}$, $\tilde{R}(X,Y)\circ S = S\circ\tilde{R}(X,Y)$, $\tilde{R}(SX,SY) = \tilde{R}(X,Y)$. Here $\underline{g}_\alpha = V_\alpha$ for $\alpha \in J$ and $\underline{g}_1 = \underline{h}^c$ (in the denotation of Lemma IV.37). Further, $\tilde{T}(V_\alpha,V_\beta) \subset V^c$ and $\tilde{R}(V_\alpha,V_\beta) \subset \underline{h}^c$ for $\alpha,\beta \in J$. Hence we get (a),(b) and (c). The last relation (d) is obtained from $[\underline{h}^c,V_\alpha] = [\underline{g}_1,\underline{g}_\alpha] \subset \underline{g}_\alpha = V_\alpha$, $\alpha \in J$. □

Let $(M,\{s_x\})$ be a connected regular s-manifold, $(V,S,\tilde{R},\tilde{T})$ its infinitesimal model. Let $J$ be the set of all eigenvalues of $S$; thus $V^c = \sum_{\alpha \in J} V_\alpha$ as in Proposition IV.38. Obviously, if $\alpha \in J$ then $\bar{\alpha} \in J$, too. $J$ is called <u>the set of eigenvalues of</u> $(M,\{s_x\})$.

From IV.38, (a), and III.42 we get the following

<u>Corollary IV.39</u>. Let $(M,g,\{s_x\})$ be a connected Riemannian regular s-manifold and $J$ its set of eigenvalues. If for every $\alpha,\beta \in J$ we have $\alpha\beta \notin J$, then $\tilde{T} = 0$, and consequently, $(M,g)$ is locally symmetric.

A subset $K \subset J$ is said to be <u>closed</u>, if

(i) $\alpha \in K$ always implies $\bar{\alpha} \in K$,

(ii) for $\alpha, \alpha' \in K$, $\alpha\alpha' \in J$ we always have $\alpha\alpha' \in K$.

<u>Theorem IV.40.</u> Let $(M,\{s_x\})$ be a connected regular s-manifold and $J$ its set of eigenvalues. Then each closed subset $K \subset J$ defines an invariant foliation of $(M,\{s_x\})$ into m-dimensional submanifolds; here $m$ is the sum of the multiplicities of all $\alpha \in K$.

<u>Proof</u>. It is sufficient to show that $K$ defines an m-dimensional invariant submanifold of $(V,S,\tilde{R},\tilde{T})$. Put $W^c = \sum_{\alpha \in K} V_\alpha$. Then $W^c$ is the complexification of an S-invariant subspace $W \subset V$, where $W$ is uniquely determined (this follows from (i)). The relation $\dim W = m$ is obvious. Now, each eigenspace $V_\alpha$ is invariant with respect to the automorphism group of $(V,S,\tilde{R},\tilde{T})$. In fact, because each automorphism commutes with $S$, it also commutes with $(S - \alpha I)^k$ and hence it preserves $V_\alpha$. Consequently, $W$ is invariant under auto-

morphisms. In particular, it follows $\tilde{R}(X,Y)(W) \subset W$ for every $X,Y \in V$. Finally, for $\alpha, \beta \in K$ we have either $\tilde{T}(V_\alpha, V_\beta) = 0$, or $\tilde{T}(V_\alpha, V_\beta) \subset V_{\alpha\beta}$, where $\alpha\beta \in K$. Hence $\tilde{T}(W,W) \subset W$ and $W$ is a submanifold of $(V,S,\tilde{R},\tilde{T})$. □

Corollary IV.41. Let $(M,g,\{s_x\})$ be a connected Riemannian regular s-manifold and $J$ its set of eigenvalues. Then each closed subset $K \subset J$ defines an invariant and totally geodesic foliation of $(M,g)$ into m-dimensional g.s. Riemannian spaces.

Proof. We only have to prove that the invariant foliation constructed as in Theorem IV.40 is totally geodesic, i.e., to check the formula from Proposition IV.8. Let $(V,g,S,\tilde{R},\tilde{T})$ be the infinitesimal model of $(M,g,\{s_x\})$. We can write $V^c = \sum_{\alpha \in J} V_\alpha = \sum_{\alpha \in K} V_\alpha \oplus \sum_{\beta \in J \setminus K} V_\beta = W^c \oplus (W^\perp)^c$. It suffices to show that

$$g^c(\tilde{T}(V_\alpha \oplus V_{\bar{\alpha}},\ V_\beta \oplus V_{\bar{\beta}}),\ V_\alpha \oplus V_{\bar{\alpha}}) = 0 \quad \text{for every } \alpha \in K,\ \beta \in J \setminus K,$$

where $g^c$ denotes the complexification of $g$. But the left-hand side can be non-zero only if $\tilde{T}(V_\alpha \oplus V_{\bar{\alpha}},\ V_\beta \oplus V_{\bar{\beta}})$ is non-zero and included in $W^c$, i.e., if $\alpha\beta \in K$ or $\alpha\bar{\beta} \in K$. Because all eigenvalues $\gamma \in J$ are now complex units ($S$ is an orthogonal transformation), it follows that $\beta$ or $\bar{\beta}$ is a product of two elements of $K$ - a contradiction to the property (ii), IV.39. □

Definition IV.42. Let $J$ be the set of eigenvalues of a regular s-manifold $(M,\{s_x\})$. $J$ is said to be reducible if there is a decomposition $J = J_1 \cup J_2$ with the following properties:

(i) if $\alpha \in J_1(J_2)$ then $\bar{\alpha} \in J_1(J_2)$ respectively,

(ii) if $\alpha, \beta, \gamma \in J$ and $\alpha\beta = \gamma$, then all $\alpha, \beta, \gamma$ belong to the same component $J_i$,

(iii) if $\alpha, \beta \in J$ and $\alpha\beta = 1$, then $\alpha, \beta$ belong to the same component $J_i$.

(Otherwise $J$ is said to be irreducible.)

Theorem IV.43. If a simply connected regular s-manifold $(M,\{s_x\})$ possesses a reducible set of eigenvalues, then it is reducible.

Proof. It is sufficient to show that the infinitesimal model $(V,S,\tilde{R},\tilde{T})$ is reducible. Clearly, both $J_1$ and $J_2$ are closed in $J$. Hence we get a direct sum decomposition $V = W_1 \oplus W_2$ of the vector space $V$, where $W_1$ and $W_2$ are invariant submanifolds of $(V,S,\tilde{R},\tilde{T})$.

It remains to show that for $Z \in W_1^c = \sum_{\alpha \in J_1} V_\alpha$, $Z' \in W_2^c = \sum_{\beta \in J_2} V_\beta$ we have $\tilde{T}(Z,Z') = 0$, $\tilde{R}(Z,Z') = 0$. But if $\alpha \in J_1$, $\beta \in J_2$ then $\alpha\beta$ is not an eigenvalue, and $\alpha\beta$ is not equal to 1. From Proposition IV.38 we get $\tilde{T}(V_\alpha, V_\beta) = 0$, $\tilde{R}(V_\alpha, V_\beta) = 0$, and the assertion follows. □

<u>Corollary IV.44.</u> Theorem IV.43 is still valid for the Riemannian case.

---

References: [F3],[H],[KN II],[K1],[K3],[LP2].

## CHAPTER V

## DISTINGUISHED s-STRUCTURES ON GENERALIZED SYMMETRIC SPACES

### Generalized affine symmetric spaces.

A connected affine manifold $(M,\tilde{\nabla})$ is called a generalized affine symmetric space (shortly, a g.a.s. space) if $M$ admits at least one regular s-structure $\{s_x\}$ such that $\tilde{\nabla}$ is its canonical connection. Each regular s-structure with this property is called admissible for $(M,\tilde{\nabla})$.

Obviously, $\{s_x\}$ is admissible if and only if all $s_x$, $x \in M$, are affine transformations with respect to $\tilde{\nabla}$ and $\tilde{\nabla}S = 0$. According to II.27, each g.a.s. space is an affine reductive space; the converse is not true. Here we give some conditions for an affine reductive space to become a g.a.s. space.

**Proposition V.1.** Let $(M,\tilde{\nabla})$ be an affine reductive space and $s: M \longrightarrow M$ an affine transformation having an isolated fixed point $o \in M$. Then $s$ comes from an admissible regular s-structure $\{s_x\}$ on $(M,\tilde{\nabla})$ if and only if $S_o = s_{*o}$ is invariant with respect to the holonomy group $\Psi(o) \subset GL(M_o)$.

Proof. If $s$ coincides with an element $s_o$ of an admissible regular s-structure $\{s_x\}$, then $S_o$ can be extended to a parallel tensor field $S$ on $(M,\tilde{\nabla})$ and hence $S_o$ is invariant with respect to the holonomy group at $o$.

Conversely, let $S_o = s_{*o}$ be invariant with respect to the holonomy group $\Psi(o)$. Denote $G = Tr(M)$, $G_o$ = the isotropy subgroup of $G$ at $o$. Then the linear isotropy representation of $G_o$ in the tangent space $M_o$ has the image $\Psi(o)$. Because $S_o$ commutes with each element of $\Psi(o)$, then $s$ commutes with each element of $G_o$. Consider the automorphism $\tilde{\sigma}: g \mapsto s \circ g \circ s^{-1}$ of $A(M)$ onto itself. Because $G$ is a normal subgroup of $A(M)$, then $\tilde{\sigma}$ induces an automorphism $\sigma$ of $G$ such that $G_o \subset (G)^{\sigma}$. Now, we can define a regular s-structure $\{s_x\}$, $x \in M$, by putting $s_x = g \circ s \circ g^{-1}$ where $g \in G$ is an arbitrary transvection satisfying $g(o) = x$. (Cf. proof of Part I, Theorem II.25.) Also, using Theorem II.25 we see that the canonical connection of $\{s_x\}$ coincides with the canonical connection $\widetilde{\nabla}$ of the reductive homogeneous space $G/G_o$. Hence $\{s_x\}$ is admissible. □

Remark V.2. For an affine transformation $s: M \longrightarrow M$ there is at most one admissible regular s-structure $\{s_x\}$ such that $s$ comes from $\{s_x\}$. It follows from the property $\widetilde{\nabla} S = 0$.

As a consequence of Proposition V.1, we obtain

Proposition V.3. Let $(M,\widetilde{\nabla})$ be a simply connected affine reductive space, and $o \in M$ a fixed point. Then the admissible regular s-structures on $(M,\widetilde{\nabla})$ are in a $(1-1)$ correspondence with the linear transformations $S_o: M_o \longrightarrow M_o$ such that

(i) $I_o - S_o$ is a non-singular transformation of $M_o$

(ii) $S_o(\widetilde{R}_o) = \widetilde{R}_o$, $S_o(\widetilde{T}_o) = \widetilde{T}_o$

(iii) $\widetilde{R}_o(S_oX,S_oY) = \widetilde{R}_o(X,Y)$ for every $X, Y \in M_o$.

Proof. (i) and (ii) mean that $S_o$ yields an affine transformation $s$ of $(M,\widetilde{\nabla})$ having $o$ as an isolated fixed point. With respect to the relation $S_o(\widetilde{R}_o) = \widetilde{R}_o$, (iii) is equivalent to the relation $\widetilde{R}_o(X,Y)\cdot S_o = 0$ for every $X, Y \in M_o$. According to the proof of Proposition I.39, this means the invariance of $S_o$ with respect to $\Psi(o)$. □

We can also state the Riemannian analogue of Proposition V.3; the proof is obvious:

Proposition V.4. Let $(M,g)$ be a simply connected generalized symmetric Riemannian space, and $o \in M$ a fixed point. Then the Riemannian regular s-structures on $(M,g)$ having a fixed canonical connection $\widetilde{\nabla}$ are in a (1-1)-correspondence with the linear transformations $S_o: M_o \longrightarrow M_o$ such that

(i) $I_o - S_o$ is a non-singular transformation of $M_o$

(ii) $S_o(\widetilde{R}_o) = \widetilde{R}_o$, $S_o(\widetilde{T}_o) = \widetilde{T}_o$, $S_o(g_o) = g_o$

(iii) $\widetilde{R}_o(S_oX,S_oY) = \widetilde{R}_o(X,Y)$ for every $X, Y \in M_o$.

-------

The symmetries of a regular s-structure $\{s_x\}$ are said to be semi-simple if the transformations $S_x: M_x \longrightarrow M_x$, $x \in M$, are semi-

-simple. For this, it is sufficient that $S_o$ be semi-simple at one fixed point $o \in M$, i.e., that $S_o$ be completely reducible on $(M_o)^c = M_o \otimes C$.

Proposition V.5. Let $\{s_x\}$ be an admissible regular s-structure on a simply connected g.a.s. space $(M,\tilde{\nabla})$. Then there is an admissible regular s-structure $\{s_x'\}$ with the same eigenvalues (including the multiplicity) as $\{s_x\}$ and such that its symmetries are semi-simple.

Proof. Consider the transformation $S_o: M_o \longrightarrow M_o$, and let $S_o = S_o' + N_o$ be the decomposition of $S_o$ into the semi-simple part and the nilpotent part. Thus, for each complex eigenspace $V_\alpha$ of $S_o$ we have $S_o'(Z) = \alpha Z$ for all $Z \in V_\alpha$. From Proposition IV.38 we can see easily that $S_o'(\tilde{T}_o) = \tilde{T}_o$, and $\tilde{\tilde{R}}_o(Z,Z')\circ S_o' = S_o'\circ\tilde{R}_o(Z,Z')$, $\tilde{R}_o(S_o'Z,S_o'Z') = \tilde{R}_o(Z,Z')$ for all complex eigenvectors $Z,Z'$ of $S_o'$. Hence the conditions (i), (ii), (iii) of Proposition V.3 follow. □

In accordance with Definition II.43, a regular s-structure $\{s_x\}$ on a manifold $M$ is said to be of order $k$ $(k \geq 2)$ if $(s_x)^k$ = identity for all $x \in M$, and if $k$ is the least number with this property. If such an integer $k$ does not exist, then $\{s_x\}$ is said to be of infinite order.

Definition V.6. A g.a.s. space $(M,\tilde{\nabla})$ is said to be of order $k$ $(k \geq 2)$ if it admits a regular s-structure of order $k$, and if it does not admit any regular s-structure of order $\ell < k$. $(M,\tilde{\nabla})$ is said to be of infinite order if it admits only s-structures of infinite order.

Definition V.7. A g.a.s. space $(M,\tilde{\nabla})$ is said to be unitary if it admits a regular s-structure $\{s_x\}$ the eigenvalues of which are complex units and the symmetries of which are semi-simple.

We have the following

Proposition V.8. A unitary g.a.s. space $(M,\tilde{\nabla})$ is always of finite order.

Proof. Let $\{s_x\}$ be an admissible regular s-structure on $(M,\tilde{\nabla})$ such that the symmetries $s_x$ are semi-simple and all eigenvalues are complex units. Let $p \in M$ be a point and denote by $Cl^a(s_p)$ the closure in $A(M,p)$ of the group generated by $s_p$. (Here $A(M,p)$ denotes the isotropy subgroup of $A(M,\tilde{\nabla})$ at the point $p$.) Let $A(M_p)$ denote the group of tangent maps at $p$ of all affine transformations $g \in A(M,p)$. Consider now the tangent map $S_p = (s_p)_{*p}: M_p \longrightarrow M_p$. Then there is a positive scalar product $B$ on $M_p$ which is invariant with respect to $S_p$. The closure $Cl(S_p)$ of the group generated by $S_p$ in the orthogonal group $O(M_p)$ is a subgroup of $A(M_p)$. According to the proof of Lemma 0.7, there is a symmetry $s'_p$ of finite order in $Cl^a(s_p)$ and the same argument as in Theorem 0.12 shows that $s'_p$ gives rise to a regular s-structure $\{s'_x\}$ as required. □

According to Theorem I.33, each g.a.s. space $(M,\tilde{\nabla})$ with a fixed origin $o$ can be represented in a unique way as a prime reductive homogeneous space. In particular, this prime reductive space is a Lie group if and only if $M$ itself is a Lie group and $\tilde{\nabla}$ is its Cartan (-)connection. Here $Tr(M) = M_{\ell}$ = the group of all left translations of $M$ (cf. I.34).

Proposition V.9. Let $K$ be a connected Lie group and $\tilde{\nabla}$ its canonical connection. If $(K,\tilde{\nabla})$ is a g.a.s. space, then $K$ is solvable.

Proof. For any admissible regular s-structure $\{s_x\}$ on $(K,\tilde{\nabla})$ we have $Tr(M,\{s_x\}) = Tr(K) \cong K$. According to I.13, $\tilde{R} = 0$ and Theorem II.41 says that $K$ is solvable. □

-------

A g.a.s. space $(M,\tilde{\nabla})$ is said to be of semi-simple type, or of solvable type, if the transvection group $Tr(M)$ is semi-simple, or solvable, respectively. Theorem IV.21 and IV.24 from Chapter IV can be now re-formulated in terms of g.a.s. spaces:

Theorem V.10. Let $(M,\tilde{\nabla})$ be a unitary g.a.s. space. Then $(M,\tilde{\nabla})$ is canonically a fibre bundle such that the base space is a g.a.s. space of semi-simple type and the fibres are autoparallel submanifolds which are g.a.s. spaces of solvable type. The connection on the base is the subjunction of $\tilde{\nabla}$.
Moreover, let the holonomy group of $(M,\tilde{\nabla})$ be compact. Then $(M,\tilde{\nabla})$ admits two complementary foliations $\mathcal{F}_1$, $\mathcal{F}_2$, which are both autoparallel; here $\mathcal{F}_1$ consists of g.a.s. spaces of semi-

-simple type and $\mathcal{F}_2$ consists of g.a.s. spaces of solvable type (all leaves are endowed with the induced connection).

Proof: we use an admissible periodic regular s-structure. □

-------

A g.a.s. space is said to be primitive if it is not a direct product of two g.a.s. spaces. According to Proposition IV.13, the direct product of g.a.s. spaces is always a g.a.s. space. On the other hand, if a g.a.s. space $(M,\tilde{\nabla})$ is a direct product of affine manifolds, then the factors are affine reductive spaces (Theorem I.37). But they need not be g.a.s. spaces; thus $(M,\tilde{\nabla})$ may be still primitive.

Example V.11: There are simply connected affine reductive spaces $(M_1,\tilde{\nabla}_1)$, $(M_2,\tilde{\nabla}_2)$ such that the direct product $(M_1\times M_2, \tilde{\nabla}_1\times\tilde{\nabla}_2)$ is a primitive g.a.s. space.

It is sufficient to construct two Lie algebras $\underline{g}_1$, $\underline{g}_2$ with the following properties:

(a) $\underline{g}_1$ does not admit an automorphism without non-zero fixed vectors,

(b) $\underline{g} = \underline{g}_1 \oplus \underline{g}_2$ admits an algebra automorphism without non-zero fixed vectors,

(c) $\underline{g} = \underline{g}_1 \oplus \underline{g}_2$ is the only direct sum decomposition of $\underline{g}$.

In fact, let $G_1$, $G_2$, $G = G_1\times G_2$ be the corresponding simply connected Lie groups and $\tilde{\nabla}_1$, $\tilde{\nabla}_2$, $\tilde{\nabla}_1\times\tilde{\nabla}_2$ their canonical connections (cf. Corollary I.36). We denote by $\mathfrak{S}$ the automorphism of $G$ corresponding to that of $\underline{g}$. Then $(G,e,\mathfrak{S})$ is a regular homogeneous s-manifold (see II.23), and $(G,\tilde{\nabla}_1\times\tilde{\nabla}_2)$ is a generalized affine symmetric space due to Theorem II.25. It is the product of affine reductive spaces $(G_1,\tilde{\nabla}_1)$, $(G_2,\tilde{\nabla}_2)$.

On the other hand, if $(G_1,\tilde{\nabla}_1)$ were a g.a.s. space, then $\tilde{\nabla}_1$ would be a canonical connection of a regular s-manifold $(G_1,\{s_x\})$. According to Theorem II.40, there would exist a prime regular homogeneous s-manifold $(\mathrm{Tr}(G_1),(\mathrm{Tr}(G_1))_o,\sigma)$, which is, in fact, of the form $((G_1)_\ell,\mathrm{id},\mathfrak{S})$. (Here we have used I.34 once again.) Consequently, $G_1$ would admit an automorphism $\sigma'$ for which $e$ is an isolated fixed point, a contradiction.

Finally, suppose that $(G,\tilde{\nabla}_1\times\tilde{\nabla}_2) = (M_1',\tilde{\nabla}_1')\times(M_2',\tilde{\nabla}_2')$ (a direct product of simply connected affine manifolds). Then $G = \mathrm{Tr}(M_1'\times M_2') = \mathrm{Tr}(M_1')\times\mathrm{Tr}(M_2')$ (cf. the proof of Theorem I.37); thus $G$ is a direct product and the only possibility for this is $Tr(M_1') = G_1$, $Tr(M_2') = G_2$ (up to a numeration). Hence $(M_i', \tilde{\nabla}_i')$ are the same as $(G_i, \tilde{\nabla}_i)$. Our product space is primitive.

Construction:

Let $\underline{g}_1$ be a non-abelian 2-dimensional Lie algebra over R. Thus, there is a basis $\{X, Y\}$ of $\underline{g}_1$ such that $[X, Y] = Y$. Let $\underline{g}_1'$ be another copy of $\underline{g}_1$ with the corresponding basis $\{X', Y'\}$. The Lie algebra $\underline{g} = \underline{g}_1 + \underline{g}_1'$ has a unique direct sum decomposition.

The linear map $S: \underline{g} \longrightarrow \underline{g}$ determined by

$$SX = X' + X, \quad SY = Y', \quad SX' = X - X', \quad SY' = -Y$$

is an automorphism without non-zero fixed vectors (the characteristic polynomial is $(\lambda^2 + 1)(\lambda^2 - 2)$).

On the other hand, consider the subalgebra $\underline{g}_1 = (X, Y)$. A map $T: \underline{g}_1 \longrightarrow \underline{g}_1$ is an automorphism if and only if $TX = X + \beta Y$, $TY = \delta Y$. Now, $(1-\delta)X + \beta Y$, or $X$, or $Y$, is a non-zero fixed vector of $T$.

## The multiplicative theory of eigenvalues.

The role of eigenvalues in the regular s-structures has been partly revealed at the end of Chapter IV. We shall now study the eigenvalues in more details. In particular, we shall show that both affine and Riemannian generalized symmetric spaces admit regular s-structures with some "distinguished" eigenvalues. We shall start with some abstract definitions.

Let $(C^n)^o$ denote the set of all n-tuples $(\theta_1, \ldots, \theta_n)$ of complex numbers such that $\theta_i \neq 0, 1$ for $i = 1, \ldots, n$. A characteristic variety of $(C^n)^o$ is a submanifold $\mathcal{V} \subset (C^n)^o$ defined by any of the following relations:

$$\left.\begin{array}{ll} \theta_i \theta_j = \theta_k & (i \neq j \neq k) \\ \theta_i \theta_j = 1 & \\ \theta_i = \bar{\theta}_j & \end{array}\right\} \; i, j, k = 1, \ldots, n \qquad (1).$$

For a permutation $\pi \in \mathfrak{S}_n$ of the indices $1, \ldots, n$ and for a subset $\mathcal{M} \subset (C^n)^o$ we put

$$I_{\pi}(\mathcal{M}) = \{(\theta_{\pi(1)},\dots,\theta_{\pi(n)}) \mid (\theta_1,\dots,\theta_n) \in \mathcal{M}\}$$

If $\mathcal{V}$ is a characteristic variety then so is $I_{\pi}(\mathcal{V})$ for each $\pi \in \mathfrak{S}_n$. Finally, we put

$$\mathfrak{S}(\mathcal{M}) = \bigcup_{\pi \in \mathfrak{S}_n} I_{\pi}(\mathcal{M}) \qquad \text{for all} \quad \mathcal{M} \subset (C^n)^o.$$

By a <u>$\theta$-variety</u> of $(C^n)^o$ we mean a non-empty set the form $\mathfrak{S}(\mathcal{V}_1 \cap \dots \cap \mathcal{V}_p)$, where $\mathcal{V}_1,\dots,\mathcal{V}_p$ are characteristic varieties. Obviously, we have only finite number of $\theta$-varieties in $(C^n)^o$; they form <u>a partially ordered set</u> with respect to the inclusion map.

Each intersection $\mathcal{V}_1 \cap \dots \cap \mathcal{V}_p$ of characteristic varieties is an algebraic variety of $(C^n)^o$ and thus it has a well-determined dimension. Hence each $\theta$-variety has a well-determined dimension.

For $(\theta_i) \in (C^n)^o$ we shall denote by $\mathcal{W}(\theta_i)$ the intersection of all characteristic varieties containing $(\theta_i)$; if there is no such characteristic variety, we put $\mathcal{W}(\theta_i) = \emptyset$. Finally, we put $\mathcal{W}^*(\theta_i) = \mathfrak{S}(\mathcal{W}(\theta_i))$. Then $\mathcal{W}^*(\theta_i)$ is the least $\theta$-variety containing $(\theta_i)$.

It is easy to see that, for each $\theta$-variety $\mathcal{W}$ of $(C^n)^o$, there is $(\theta_i) \in \mathcal{W}$ such that $\mathcal{W} = \mathcal{W}^*(\theta_i)$. In fact, let us suppose that $\{\mathcal{V}_1,\dots,\mathcal{V}_p\}$ is a maximal set of characteristic varieties for which $\mathcal{W} = \mathfrak{S}(\mathcal{V}_1 \cap \dots \cap \mathcal{V}_p)$. For any characteristic variety $\mathcal{V} \neq \mathcal{V}_1,\dots,\mathcal{V}_p$, the intersection $\mathcal{V} \cap \mathcal{W}$ has dimension less than $\mathcal{W}$. If $\mathcal{V}_{p+1},\dots,\mathcal{V}_s$ are all remaining characteristic varieties of $(C^n)^o$ then $\mathcal{W} \cap (\mathcal{V}_{p+1} \cup \dots \cup \mathcal{V}_s)$ is a set of measure zero in $\mathcal{W}$. Now, it is sufficient to choose $(\theta_i) \in \mathcal{W} \setminus (\mathcal{V}_{p+1} \cup \dots \cup \mathcal{V}_s)$.

Now, let $\mathcal{A}^n$ denote the subset of all elements $(\theta_i) \in (C^n)^o$ with the following property: if $\theta$ is among $\theta_1,\dots,\theta_n$ with the multiplicity m then so is its conjugate $\bar{\theta}$. Equivalently: $(\theta_i) \in (C^n)^o$ belongs to $\mathcal{A}^n$ if and only if there is a permutation $\wp \in \mathfrak{S}_n$ such that $\wp^2 = \mathrm{id}$ and $\bar{\theta}_i = \theta_{\wp(i)}$ for $i = 1,\dots,n$. From (1) we see that $\mathcal{A}^n$ is a union of certain $\theta$-varieties of $(C^n)^o$. More precisely, we can prove the following:

<u>Lemma V.12.</u> For $(\theta_i) \in \mathcal{A}^n$ we always have $\mathcal{W}^*(\theta_i) \subset \mathcal{A}^n$.

<u>Proof.</u> Because $\mathcal{W}^*(\theta_i) = \mathfrak{S}(\mathcal{W}(\theta_i))$, and $\mathfrak{S}(\mathcal{A}^n) = \mathcal{A}^n$, it is sufficient to prove $\mathcal{W}(\theta_i) \subset \mathcal{A}^n$. Suppose $(\theta_i) \in \mathcal{A}^n$, and let $\wp \in \mathfrak{S}_n$ be a permutation such that $\wp^2 = \mathrm{id}$, and $\bar{\theta}_i = \theta_{\wp(i)}$ for $i = 1,\dots,n$. In other words, $\mathcal{W}(\theta_i) \subset \mathcal{W}_1 \cap \dots \cap \mathcal{W}_n$, where $\mathcal{W}_i$ denotes the characteristic variety of $(C^n)^o$ given by the relation

$\bar{\theta}_i = \theta_{\varphi(i)}$. For $(\theta'_i) \in \mathcal{W}(\theta_i)$ we get $\bar{\theta}'_i = \theta'_{\varphi(i)}$ for $i = 1,\dots,n$ and hence $(\theta'_i) \in \mathcal{A}^n$. □

As a consequence, we have the following

**Proposition V.13.** Each minimal $\theta$-variety $\mathcal{W} \subset (C^n)^o$ either belongs to $\mathcal{A}^n$ or to the complement $(C^n)^o \setminus \mathcal{A}^n$.

**Lemma V.14.** If $(\theta_i) \in \mathcal{A}^n$ and $(\theta'_i) \in \mathcal{W}(\theta_i)$, then every equality between the numbers $\theta_i$ implies the equality between the corresponding numbers $\theta'_i$.

Proof. Suppose that $\theta_j = \theta_k$ for some $j,k$ where $j \neq k$. Because $(\theta_i) \in \mathcal{A}^n$, there is a permutation $\varphi \in \mathfrak{S}_n$ such that $\theta_{\varphi(j)} = \bar{\theta}_j$, $\theta_{\varphi(k)} = \bar{\theta}_k$. Hence $\theta_{\varphi(j)} = \bar{\theta}_k$, $\theta_{\varphi(k)} = \bar{\theta}_j$. Because $(\theta'_i) \in \mathcal{W}(\theta_i)$, we obtain the same characteristic relations for $\theta'_1,\dots,\theta'_n$ and hence $\theta'_j = \theta'_k$.

-------

Let now $(M,\{s_x\})$ be a regular s-manifold, $\dim M = n$, and $(V,S,\tilde{R},\tilde{T})$ its infinitesimal model. An n-tuple $(\theta_1,\dots,\theta_n) \in (C^n)^o$ is called a system of eigenvalues of $\{s_x\}$ if all $\theta_i$ are eigenvalues of the linear transformation $S: V \longrightarrow V$ and each eigenvalue $\alpha$ of $S$ occurs in $(\theta_1,\dots,\theta_n)$ with the corresponding multiplicity. Hence a system of eigenvalues of $\{s_x\}$ is uniquely determined exactly up to a permutation of its terms.

Because $S: V \longrightarrow V$ is a non-singular real transformation and $I - S$ is also non-singular, we always have $(\theta_i) \in \mathcal{A}^n$.

**Proposition V.15.** Let $(M,\tilde{\nabla})$ be a simply connected g.a.s. space. If $(M,\tilde{\nabla})$ admits a regular s-structure $\{s_x\}$ with a system of eigenvalues $(\theta_i)$, and if $(\theta'_j) \in \mathcal{W}^*(\theta_i)$, then $(M,\tilde{\nabla})$ admits a regular s-structure $\{s'_x\}$ with the system of eigenvalues $(\theta'_j)$.

Proof. Let $(V,S,\tilde{R},\tilde{T})$ be the infinitesimal model of $(M,\{s_x\})$. According to Proposition V.5 we can suppose that $S: V \longrightarrow V$ is completely reducible. Denote by $\{U_1,\dots,U_n\}$ a basis of complex eigenvectors for $S$ corresponding to the eigenvalues $\theta_1,\dots,\theta_n$ respectively. Moreover, we can suppose that a permutation $\varphi \in \mathfrak{S}_n$ exists such that $\varphi^2 = \mathrm{id}$, $\bar{\theta}_i = \theta_{\varphi(i)}$, $\bar{U}_i = U_{\varphi(i)}$ for $i = 1,\dots,n$.

Supposing $(\theta'_j) \in \mathcal{W}^*(\theta_i)$, we find a permutation $\pi \in \mathfrak{S}_n$ such that $I_\pi(\theta'_i) \in \mathcal{W}(\theta_i)$. Thus, we can re-order the numbers $\theta'_1,\dots,\theta'_n$ in such a way that $(\theta'_i) \in \mathcal{W}(\theta_i)$. This means that each characteristic relation satisfied by $\theta_1,\dots,\theta_n$ is also satisfied by $\theta'_1,\dots,\theta'_n$.

Now, let us define a linear transformation $S'$ of $V^c = V \otimes C$ by the relations $S'U_i = \theta'_i U_i$, $i = 1,\dots,n$. We have to show that $S'$ induces a real transformation of $V$ and that it satisfies conditions (i), (ii), (iii) of Proposition V.4.

Firstly, we have $\bar{\theta}'_i = \theta'_{\varphi(i)}$ for each $i$, and hence $S'U_i = \theta'_i U_i$ implies $S'\bar{U}_i = S'U_{\varphi(i)} = \theta'_{\varphi(i)}U_{\varphi(i)} = \bar{\theta}'_i\bar{U}_i$; thus $S'$ comes from a real transformation on $V$. Further, let $\tilde{T}(U_i,U_j) = \sum_k T^k_{ij}U_k$. Because $S(\tilde{T}) = \tilde{T}$, i.e., $S(\tilde{T}(U_i,U_j)) = \tilde{T}(SU_i,SU_j)$, we can have $T^k_{ij} \neq 0$ only if $\theta_i\theta_j = \theta_k$, or else, only if $\theta'_i\theta'_j = \theta'_k$. Hence $\tilde{T}(S'U_i,S'U_j) = \theta'_i\theta'_j\tilde{T}(U_i,U_j) = \theta'_i\theta'_j(\sum_k T^k_{ij}U_k) = \sum_k T^k_{ij}\theta'_kU_k = \sum_k T^k_{ij}S'U_k = S'(\tilde{T}(U_i,U_j))$, i.e., $S'(\tilde{T}) = \tilde{T}$.

Further, we have $\tilde{R}(SU_i,SU_j) = \tilde{R}(U_i,U_j)$ for every $i$, $j$. Hence $\tilde{R}(U_i,U_j) \neq 0$ only if $\theta_i\theta_j = 1$, or else, only if $\theta'_i\theta'_j = 1$. We get easily $\tilde{R}(S'U_i,S'U_j) = \tilde{R}(U_i,U_j)$.

Now, let $A$ be an endomorphism such that $A(S) = 0$. Then $A$ leaves invariant the eigenspaces of $S$ corresponding to the mutually different eigenvalues among $\theta_1,\dots,\theta_n$. According to Lemma V.14, each maximal eigenspace of $S'$ is a direct sum of certain maximal eigenspaces of $S$. Consequently, $A$ leaves invariant the maximal eigenspaces of $S'$ and thus $A(S') = 0$. In particular, $\tilde{R}(X,Y)(S) = 0$ implies $\tilde{R}(X,Y)(S') = 0$. Then the relations $\tilde{R}(S'X,S'Y) = \tilde{R}(X,Y)$ and $\tilde{R}(X,Y)(S') = 0$ yield $\tilde{R}(S'X,S'Y)S'Z = S'(\tilde{R}(X,Y)Z)$, i.e., $S'(\tilde{R}) = \tilde{R}$. This completes the proof. □

Now, consider the partially ordered set (with respect to the inclusion map) of all $\theta$-varieties of $(C^n)^o$, and let $\mathfrak{L}^n$ denote the union of all minimal $\theta$-varieties which are contained in $\mathcal{A}^n$ (cf. Proposition V.13). We obtain from Proposition V.15 the following

Theorem V.16. Each simply connected g.a.s. space $(M,\tilde{\nabla})$ admits a regular s-structure such that its system of eigenvalues belongs to $\mathfrak{L}^n$.

Denote by $\mathfrak{B}^n$ the subset of all n-tuples $(\theta_1,\dots,\theta_n) \in (C^n)^o$ such that $|\theta_1| = \dots = |\theta_n| = 1$.

Lemma V.17. If $(\theta_i) \in \mathcal{A}^n \cap \mathfrak{B}^n$, then $\mathcal{W}^*(\theta_i) \subset \mathcal{A}^n \cap \mathfrak{B}^n$.

Proof. Suppose $(\theta_i) \in \mathcal{A}^n \cap \mathcal{B}^n$, $(\theta'_j) \in \mathcal{W}^*(\theta_i)$, and let us re-order the numbers $\theta'_j$ in such a way that $(\theta'_i) \in \mathcal{W}(\theta_i)$. Because $(\theta_i) \in \mathcal{A}^n$, there is a permutation $\wp \in \mathfrak{S}_n$ such that $\bar{\theta}_i = \theta_{\wp(i)}$ for $i = 1,\ldots,n$. Now $|\theta_i| = 1$ means that $\theta_i \theta_{\wp(i)} = 1$ and hence $\theta'_i \theta'_{\wp(i)} = 1$ for $i = 1,\ldots,n$. Because it holds also $\bar{\theta}'_i = \theta'_{\wp(i)}$, we get $|\theta'_i| = 1$ for $i = 1,\ldots,n$. Hence $(\theta'_j) \in \mathcal{A}^n \cap \mathcal{B}^n$. □

As a consequence we get

**Proposition V.18.** Each minimal $\theta$-variety $\mathcal{W} \subset (C^n)^o$ either belongs to $\mathcal{A}^n \cap \mathcal{B}^n$ or to its complement in $(C^n)^o$.

Let $\mathcal{D}^n$ denote the union of all minimal $\theta$-varieties which are contained in $\mathcal{A}^n \cap \mathcal{B}^n$. We obtain from Definition V.7 and Proposition V.15:

**Theorem V.19.** Each simply connected unitary g.a.s. space $(M,\tilde{\nabla})$ admits a regular s-structure such that its system of eigenvalues belongs to $\mathcal{D}^n$.

We can now state an analogous theorem for generalized symmetric Riemannian spaces.

**Theorem V.20.** Each simply connected generalized symmetric Riemannian space $(M,g)$ admits a regular s-structure such that its system of eigenvalues belongs to $\mathcal{D}^n$.

Proof. Let $\{s_x\}$ be an admissible s-structure of $(M,g)$ and $\tilde{\nabla}$ its canonical connection. Then the system of eigenvalues of $\{s_x\}$ belongs to $\mathcal{A}^n \cap \mathcal{B}^n$; hence $(M,\tilde{\nabla})$ is a unitary g.a.s. space. According to the proof of Proposition V.15 (and taking into account Proposition V.4) we can see easily that, for every $(\theta'_j) \in \mathcal{W}^*(\theta_i)$, the space $(M,g)$ admits a regular s-structure $\{s'_x\}$ with the system of eigenvalues $(\theta'_j)$. Hence the result follows. □

## The additive theory of eigenvalues.

The aim of this paragraph is to clear up the structure of the set $\mathcal{D}^n$. For this purpose, we shall develop the "additive" theory of the set $\mathcal{B}^n$ and of its $\theta$-varieties. Consider the diffeomorphism $f$ of the open unit cube $(I^n)^o \subset R^n$ onto $\mathcal{B}^n$ given as follows:

$$f([x^1,\dots,x^n]) = (\exp(2\pi x^1\sqrt{-1}),\dots,\exp(2\pi x^n\sqrt{-1})) \qquad (2).$$

By means of $f$, each $\Theta$-variety $\mathcal{W} \subset \mathcal{B}^n$ can be represented as a finite "bunch" of linear subspaces of $R^n$ having non-empty intersections with $(I^n)^o$. Thus we can describe the structure of $\mathcal{D}^n$ in terms of linear subspaces of $R^n$.

We shall start with a geometric result called the "Basic Lemma". Let $\Lambda_n$ be a finite set of linear subspaces of the cartesian space $R^n[x^1,\dots,x^n]$ with the following properties:

(a) For $K, L \in \Lambda_n$ we have $K \cap L \in \Lambda_n$,

(b) $\Lambda_n$ contains all hyperplanes given by the equations of the form $x^i - x^j = 0$, or $x^i + x^j = 1$, where $1 \leq i < j \leq n$.

Let $(I^n)^o$ denote the open unit cube

$$(I^n)^o = \{[x^1,\dots,x^n] \in R^n \mid 0 < x^i < 1, \quad i = 1,\dots,n\}.$$

$\mathcal{L}_n \subset \Lambda_n$ will denote the subset of all 0-dimensional subspaces; they will be called <u>the lattice points</u> of $\Lambda_n$.

<u>Basic Lemma V.21.</u> Each linear subspace $K \in \Lambda_n$ of dimension $k > 0$ such that $K \cap (I^n)^o \neq \emptyset$ contains a proper subspace $L \in \Lambda_n$ of dimension $\ell < k$ such that $L \cap (I^n)^o \neq \emptyset$. In particular, if $K \in \Lambda_n$, $K \cap (I^n)^o \neq \emptyset$, then $K \cap (I^n)^o$ contains a lattice point $p \in \mathcal{L}_n$.

First, we shall need some other lemmas.

<u>Lemma V.22.</u> (Generalized Pasch's axiom.) Let $\Delta^{(r)}$ be an r-dimensional simplex in the euclidean space $E^r$, and let a linear subspace $K \subset E^r$ with $\dim K > 0$ intersect the interior of $\Delta^{(r)}$. Then $K$ intersects the interior of a (r-1)-dimensional face $\Delta_i^{(r-1)}$ and the interior of an $\ell$-dimensional face $\Delta_j^{(\ell)}$ $(\ell \leq r-1)$ such that $\Delta_j^{(\ell)} \not\subseteq \Delta_i^{(r-1)}$.

<u>The proof</u> is left to the reader.

Let $I^n$ $(n \geq 2)$ denote the closed n-dimensional unit cube

$$I^n = \{[x^1,\dots,x^n] \in R^n \mid 0 \leq x^i \leq 1, \quad i = 1,\dots,n\}.$$

<u>Lemma V.23</u>. $I^n$ admits a triangulation into $2^{n-1}\cdot n!$ simplexes $\Delta_i^{(n)}$ of dimension $n$ such that

(a) Each (n-1)-dimensional face of each simplex $\Delta_i^{(n)}$ is contained in one of the following hyperplanes:

$$x^i = 0, \quad \text{or} \quad x^i = 1, \quad \text{or} \quad x^i - x^j = 0 \quad (i \neq j), \\ \text{or} \quad x^i + x^j = 1 \quad (i \neq j), \quad \text{where} \quad i,j = 1,\dots,n \qquad (3).$$

(b) Exactly one (n-1)-dimensional face of each simplex $\Delta_i^{(n)}$ belongs to the boundary $\partial I^n$ of $I^n$.

(c) The center of the cube $I^n$ is the common vertex of all simplexes $\Delta_i^{(n)}$.

<u>Proof</u>. For $n = 2$, the required triangulation of $I^2$ is given by means of the 6 lines $x^1 = 0$, $x^2 = 0$, $x^1 = 1$, $x^2 = 1$, $x^1 - x^2 = 0$, $x^1 + x^2 = 1$. Suppose Lemma V.23 to be true for some $n$, and consider the cube $I^{n+1}$. Let $T$ be a triangulation of $I^n$ satisfying the conditions of Lemma V.20, and let $f_{i,\alpha}: I^n \longrightarrow I^{n+1}$ $(i = 1,\dots,n+1;$ $\alpha = 0,1)$ denote the map

$$f_{i,\alpha}(x^1,\dots,x^n) = [x^1,\dots,x^{i-1},\alpha,x^i,\dots,x^n].$$

Then we get a triangulation $f_{i,\alpha}(T)$ on each face $f_{i,\alpha}(I^n)$ of $I^{n+1}$, and thus a triangulation of the boundary $\partial(I^{n+1})$. Now, we define a triangulation $T'$ of $I^{n+1}$ in such a way that each (n+1)-dimensional simplex $\Delta_i^{(n+1)}$ of $T'$ has the center $[\frac{1}{2},\dots,\frac{1}{2}]$ of $I^{n+1}$ for a vertex and an n-dimensional simplex of the boundary triangulation for a face. It is obvious that $T'$ consists of $2(n+1)\cdot 2^{n-1}\cdot n! = 2^n\cdot(n+1)!$ simplexes $\Delta_i^{(n+1)}$, and that the conditions (b), (c) are also satisfied. The verification of the condition (a) is left to the reader. □

<u>Lemma V.24</u>. Let $T$ be a triangulation of the cube $I^n$ satisfying the conditions of Lemma V.23. Further, let $\Lambda_n$ be a set of linear subspaces of $R^n$ as in the Basic Lemma. Then each k-dimensional simplex $\Delta_j^{(k)}$ $(k \leq n-1)$ of the triangulation $T$ is either contained in the boundary $\partial I^n$, or it is contained in a linear subspace $K \in \Lambda_n$ and has the point $[\frac{1}{2},\dots,\frac{1}{2}]$ for a vertex.

<u>Proof</u>. Each k-dimensional simplex $\Delta_i^{(k)}$ of $T$ is the intersection of certain number of (n-1)-dimensional simplexes of $T$. Thus the

k-dimensional plane $L$ containing $\Delta_i^{(k)}$ is determined by a system of $n-k$ equations of the form (3). Now, if $\Delta_i^{(k)}$ is not contained in the boundary $\partial I^n$, it has the point $[\frac{1}{2},\ldots,\frac{1}{2}]$ for a vertex. Hence $L$ is given by a system of equations of the form $x^i - x^j = 0$, $x^\ell + x^m = 1$, $(\ell \neq m)$, and thus it belongs to $\Lambda_n$. □

Proof of Basic Lemma. Let $T$ be the triangulation of $I^n$ constructed in Lemma V.23. Let $K \in \Lambda_n$ be of dimension $k > 0$ and such that $K \cap (I^n)^o \neq \emptyset$. Let $\Delta_i^{(r)} \in T$ be a simplex of the minimum dimension containing a point of $K \cap (I^n)^o$ in its interior. Let $L \supset \Delta_i^{(r)}$ be the corresponding r-dimensional plane. According to Lemma V.24, we have $L \in \Lambda_n$, and $[\frac{1}{2},\ldots,\frac{1}{2}]$ is a vertex of $\Delta_i^{(r)}$.

Suppose now $K \subseteq L$. Then, according to Lemma V.22, the k-dimensional plane $K$ intersects the interior of a face $\Delta_u^{(r-1)}$ and the interior of a face $\Delta_v^{(\ell)}$, where $\ell \leq r-1$ and $\Delta_v^{(\ell)} \not\subset \Delta_u^{(r-1)}$. Then at least one of the simplexes $\Delta_u^{(r-1)}$, $\Delta_v^{(\ell)} \in T$ contains the vertex $[\frac{1}{2},\ldots,\frac{1}{2}]$ and thus it does not belong to the boundary $\partial I^n$. Moreover, it contains in its interior a point $q \in K \cap (I^n)^o$ - a contradiction to the minimality of $r$.

Hence $K \not\subset L$, and the subspace $P = K \cap L$ is a proper subspace of $K$. We have $P \in \Lambda_n$ and $P \cap (I^n)^o \neq \emptyset$, which completes the proof. □□

Theorem V.25. The union $\mathcal{D}^n$ of all minimal $\theta$-varieties of $\mathcal{A}^n \cap \mathcal{B}^n$ is a finite set. Moreover, each element $(\theta_i) \in \mathcal{D}^n$ is of finite order, i.e., $(\theta_1)^k = (\theta_2)^k = \ldots = (\theta_n)^k = 1$ for some integer $k$.

Proof. Consider the set of all hyperplanes of $R^n$ which correspond to the following linear equations:

$$\left.\begin{aligned} x^i + x^j - x^k &= 0 \\ x^i + x^j - x^k &= 1 \end{aligned}\right\rangle (i \neq j \neq k,\quad i,j,k = 1,\ldots,n)$$

$$\left.\begin{aligned} x^i + x^j &= 1 \\ x^i - x^j &= 0 \end{aligned}\right\rangle (i \neq j,\quad i,j = 1,\ldots,n) \qquad (4)$$

$$2x^i = 1 \qquad (i = 1,\ldots,n).$$

Let $\Lambda_n$ denote the set of all linear subspaces of $R^n$ which are intersections of finite numbers of hyperplanes given above. For each

$L \in \Lambda_n$ we shall take into consideration all hyperplanes of the form (4) containing L. Thus, each subspace $L \in \Lambda_n$ is characterized by a unique (maximal) set $\mu(L)$ of linear non-homogeneous equations of the form (4); the equations of this set may be linearly dependent. For each subspace L we also consider the corresponding set h(L) of linear homogeneous equations. Now, two subspaces $L, L' \in \Lambda_n$ will be said to be <u>conjugate</u> if $h(L) = h(L')$.

> <u>Lemma V.26.</u> Let $L \in \Lambda_n$ be such that $L \cap (I^n)^o \neq \emptyset$ and $f(L \cap (I^n)^o) \subset \mathcal{A}^n \cap \mathcal{B}^n$. Then all equations of the form $x^i - x^j = 0$ in the set $\mu(L)$ are consequences of the equations of the form $x^k + x^\ell = 1$ in $\mu(L)$.

<u>Proof.</u> Let $[a^1,\ldots,a^n] \in L \cap (I^n)^o$ be a general point, i.e. a point not belonging to a linear subspace $L' \in \Lambda_n$ such that $\dim L' < \dim L$. Put $(\theta_1,\ldots,\theta_n) = f([a^1,\ldots,a^n])$. Then $(\theta_i) \in \mathcal{A}^n$ and there is a permutation $\wp$ of the indices $1,\ldots,n$ such that $\bar{\theta}_i = \theta_{\wp(i)}$ for $i = 1,\ldots,n$. Hence $a^i + a^{\wp(i)} = 1$ for all i. Because $[a^1,\ldots,a^n]$ was a general point of L, we conclude that $\mu(L)$ contains all the relations $x^i + x^{\wp(i)} = 1$, $i = 1,\ldots,n$. Thus the set $\mu(L)$ contains with every equation $x^i - x^j = 0$ also the equation $x^j + x^{\wp(i)} = 1$. Hence the result follows. □

For any permutation $\pi \in \mathfrak{S}_n$ consider the transformation $I_\pi : [x^1,\ldots,x^n] \mapsto [x^{\pi(1)},\ldots,x^{\pi(n)}]$ of $R^n$, and for $L \subset R^n$ put $\mathfrak{S}(L) = \bigcup_{\pi \in \mathfrak{S}_n} I_\pi(L)$.

> <u>Lemma V.27.</u> If $\mathcal{W} \subset \mathcal{B}^n$ is a $\theta$-variety then $f^{-1}(\mathcal{W}) = \mathfrak{S}(L_1 \cup \ldots \cup L_r) \cap (I^n)^o$, where $L_1,\ldots,L_r \in \Lambda_n$ form a complete set of mutually conjugate subspaces. Conversely, if $L_1,\ldots,L_r \in \Lambda_n$ is a complete set of mutually conjugate subspaces and $f(L_\alpha \cap (I^n)^o) \subset \mathcal{A}^n \cap \mathcal{B}^n$ for $\alpha = 1,\ldots,r$, then $f(\mathfrak{S}(L_1 \cup \ldots \cup L_r) \cap (I^n)^o)$ is either an empty set or a $\theta$-variety.

<u>Proof.</u> Substituing $\theta_i = \exp(2\pi x^i\sqrt{-1})$, where $x^i \in (0, 1)$ for $i = 1,\ldots,n$, into (1), we obtain only relations of the form (4) and hence the first assertion follows. Conversely, let $L_1,\ldots,L_r \in \Lambda_n$ be a complete set of mutually conjugate subspaces and suppose

$f[\mathfrak{S}(L_1 \cup \ldots \cup L_r) \cap (I^n)^o] \neq \emptyset$, $f(L_\alpha \cap (I^n)^o) \subset \mathcal{A}^n \cap \mathcal{B}^n$ for $\alpha = 1,\ldots,r$. It suffices to show that $f[(L_1 \cup \ldots \cup L_r) \cup (I^n)^o]$ is an intersection of characteristic varieties (1). Choose $\alpha \in \{1,\ldots,r\}$ such that $L_\alpha \cap (I^n)^o \neq \emptyset$. From Lemma V.26 it follows that $L_\alpha$ is an intersection of hyperplanes of the form $x^i + x^j - x^k = 0$ (or $= 1$) and of the form $x^\ell + x^m = 1$. Then $f(L_\alpha \cap (I^n)^o)$ is an open subset of the intersection of the corresponding characteristic varieties $\theta_i \theta_j = \theta_k$, $\theta_\ell \theta_m = 1$ and $\mathcal{B}^n$, or else, it is an open subset of an intersection of characteristic varieties of the form $\theta_i \theta_j = \theta_k$, $\theta_\ell \theta_m = 1$, $\theta_\ell = \bar{\theta}_m$. Because the conjugate set $\{L_1,\ldots,L_r\}$ is complete, we see that $f[(L_1 \cup \ldots \cup L_r) \cap (I^n)^o]$ coincides with this intersection. □

**Lemma V.28.** If $\mathcal{W} \subset \mathcal{A}^n \cap \mathcal{B}^n$ is a minimal $\theta$-variety then $f^{-1}(\mathcal{W}) \subset \mathcal{L}_n$.

Proof. Let $\mathcal{W} \subset \mathcal{A}^n \cap \mathcal{B}^n$ be a $\theta$-variety and put $f^{-1}(\mathcal{W}) = \mathfrak{S}(L_1 \cup \ldots \cup L_r) \cap (I^n)^o$. Suppose that $\dim L_1 > 0$; $L_1 \cap (I^n)^o \neq \emptyset$. The set $\Lambda_n$ satisfies the conditions of Basic Lemma. Thus, there is a lattice point $p \in L_1 \cap (I^n)^o$. Obviously, the set $L_1 \cup \ldots \cup L_r$ contains all lattice points $p_1,\ldots,p_s$ which are conjugate to $p$, and $f(\mathfrak{S}\{p_1,\ldots,p_s\} \cap (I^n)^o) \subset \mathcal{W}$ is a $\theta$-variety (see Lemma V.27). Thus $\mathcal{W}$ cannot be minimal, and our Lemma follows. □

Proof of Theorem V.25, continuation: According to Lemma V.28, there is a (1 - 1) correspondence between the set $\mathfrak{D}^n$ and the set $\mathcal{L}_n \cap (I^n)^o$. Hence $\mathfrak{D}^n$ is finite. Now, all lattice points of $\Lambda_n$ have rational coordinates, and consequently, the points $(\theta_i)$ of $\mathfrak{D}^n$ are elements of finite order. □ □

In addition to Theorem V.25 we shall prove the following

**Theorem V.29.** Let $k(n)$ denote the maximum of orders of all elements of $\mathfrak{D}^n$. Then $k(n) \leq 5^{n/4}$ for $n$ even, and $k(n) \leq 2\cdot 5^{(n-1)/4}$ for $n$ odd.

Let $[a^1,\ldots,a^n] \in (I^n)^o$ be a lattice point of $\Lambda_n$. If $(\theta_1,\ldots,\theta_n) = f([a^1,\ldots,a^n]) \in \mathcal{A}^n$, then there is a permutation $\wp \in \mathfrak{S}_n$ such that $\wp^2 =$ identity, and $\bar{\theta}_i = \theta_{\wp(i)}$, i.e., $a^i + a^{\wp(i)} = 1$ for $i = 1,\ldots,n$. Suppose $\wp(i) \neq i$ for $i = 1,\ldots,2r$ and $\wp(j) = j$ for $j = 2r+1,\ldots,n$. We can achieve by a re-numeration that $0 <$

$< a^i < \frac{1}{2}$ for $i = 1,\dots,r$ and $\wp(i) = r + i$ for $i = 1,\dots,r$. Naturally, we have $a^{2r+1} = \dots = a^n = \frac{1}{2}$. Put $b^i = a^i$ for $i = 1,\dots,r$, and $b^j = a^{r+j}$ for $j = r + 1,\dots,n - r$. Now, the lattice point $[a^1,\dots,a^n]$ can be calculated from the values $b^1,\dots,b^{n-r}$. On the other hand, the numbers $a^1,\dots,a^n$ are uniquely determined by the set $\mu([a^i])$ of linear equations of the form (4). Thus the corresponding values $b^1,\dots,b^r,b^{r+1},\dots,b^{n-r}$ are uniquely determined by a set of equations of the form

$$\left.\begin{array}{ll} y^i + y^j - y^k = 0 & \\ & (i \neq j \neq k; \quad i,j,k = 1,\dots,n-r) \\ y^i + y^j + y^k = 1 & \\ 2y^i - y^k = 0, \quad 2y^i + y^k = 1, \quad y^i - y^k = 0 & \\ & (i \neq k, \quad i,k = 1,\dots,n-r) \\ 2y^i = 1 & (i = r+1,\dots,n-r)\ . \end{array}\right\} \qquad (5)$$

This means that we can select a system of $n-r$ independent equations of the form (5) and then calculate $b^1,\dots,b^{n-r}$ using the Cramer's rule. Let $\Delta$ be the matrix of the left-hand side of this system, and let $|\Delta|$ denote the absolute value of $\det\Delta$. Obviously, if we find an upper bound for $|\Delta|$, we get an upper bound for the order of $(\theta_i)$.

Now, each matrix $\Delta$ belongs to a special class of matrices which is defined as follows: first, define the <u>weight</u> of a row (or column) of a square matrix $\Delta$ as the sum of the absolute values of all elements of this row (or column). A square matrix $\Delta$ is said to be <u>of type $\sigma$</u> if its elements are integers satisfying $|a_i^j| \leq 2$, and the weight of each row is $\leq 3$. For an n-matrix $\Delta_n$ of type $\sigma$ we obviously have $|\Delta_n| \leq 3^n$. More precisely, we can prove

<u>Lemma V.30.</u> For an arbitrary matrix $\Delta_n$ of type $\sigma$ with n rows we have

$$|\Delta_n| \leq 5^{n/2} \qquad (6).$$

<u>The proof</u> is rather technical and it will be omitted (see [K9], Lemma 7).

<u>Proof of Theorem V.29.</u> Consider a given $(\theta_i) \in \mathfrak{D}^n$ and the lattice point $[a^1,\dots,a^n] = f^{-1}((\theta_i))$. Then we can find a compatible system

of n-r independent equations of the form (5) defining the corresponding values $b^1,\dots,b^r,b^{r+1},\dots,b^{n-r}$. Without the loss of generality we can suppose that all equations of the form $2y^j = 1$ $(j = r+1,\dots$ $\dots,n-r)$ form a part of this system. The matrix of the left hand side of the system takes on the form

$$D = \left|\begin{array}{c:c} \Delta_r & * \\ \hdashline 0 & \begin{matrix} 2 & & \\ & \ddots & \\ & & 2 \end{matrix} \end{array}\right|$$

where $\Delta_r$ is a matrix of type $\sigma$ with r rows and columns. Solving the systems by means of the Cramer's rule, we obtain for $i = 1,\dots$ $\dots,r$: $|y^i| = |D_i|/(2^{n-2r}\cdot|\Delta_r|)$, where $D_i$ is the matrix obtained by replacing the i-th column of D by a column consisting of numbers 0 and $\pm 1$. Clearly, if $n - 2r > 0$, then $|D_i| = 2^{n-2r-1}\cdot n_i$, where $n_i$ is an integer. Hence $b^i = n_i/(2|\Delta_r|)$ for $i = 1,\dots,r$ and $b^j = 1/2$ for $j = r+1,\dots,n-r$. Thus, if $n - 2r > 0$, then $2|\Delta_r|\cdot b^i$ is an integer for $i = 1,\dots,n-r$, and $2|\Delta_r|\,a^j$ is also an integer for $j = 1,\dots,n$. If $n - 2r = 0$, then $|\Delta_r|b^i$ is an integer for $i = 1,\dots,r$ and also $|\Delta_r|a^j$ is an integer for $j = 1,\dots,n$.

Now, if n is odd, then $n - 2r > 0$, and $(\theta_i)$ is of order $k \leq 2|\Delta_r|$, where $|\Delta_r| \leq 5^{r/2} \leq 5^{(n-1)/4}$ according to Lemma V.30. If n is even, then either $(\theta_i)$ is of order $k \leq |\Delta_r|$, where $r = n/2$, or of order $k \leq 2|\Delta_r|$, where $r \leq (n-2)/2$. In both cases, $k \leq 5^{n/4}$. This completes the proof. □ □

## Applications and problems.

Applications to our theory, which follow immediately from Theorems V.19, V.20, V.25, V.29, are the following:

<u>Theorem V.31.</u> Let $(M,g)$ be a simply connected generalized symmetric Riemannian space of dimension n, and let k denote its order. Then $k \leq 5^{n/4}$ for n even, and $k \leq 2\cdot 5^{(n-1)/4}$ for n odd.

<u>Theorem V.32.</u> Let $(M,\tilde{\nabla})$ be a simply connected unitary g.a.s. space of dimension n, and let k denote its order. Then $k \leq 5^{n/4}$ for n even and $k \leq 2\cdot 5^{(n-1)/4}$ for n odd.

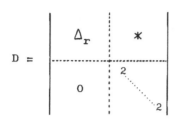

The ideas developed in this Chapter can be also used in more general situations, namely for the study of so called "periodic" tensor structures. (Such tensor structures arise naturally in the Riemannian geometry.)

Let $V$ be a real vector space and $T_1,\dots,T_r$ tensors on $V$ of types $(1,n_1),\dots,(1,n_r)$ respectively. (Here "of type $(m,n)$" means m-times contravariant and n-times covariant.) It is obvious how to define an automorphism of the tensor structure $\mathcal{T} = (V;\ T_1,\dots,T_r)$. Now, $\mathcal{T}$ is said to be periodic if it admits a periodic automorphism without non-zero fixed vectors. Order $k(\mathcal{T})$ of a periodic $\mathcal{T}$ is defined as the minimum order of such an automorphism.

Now, let $\Pi_n^{n_1,\dots,n_r}$ denote the class of all periodic tensor structures $(V;\ T_1,\dots,T_r)$ where $n_1,\dots,n_r$ are fixed and $\dim V = n$, and let $\tilde{\Pi}_n^{n_1,\dots,n_r}$ denote some of its subclasses. (Such a subclass is defined by the additional conditions put on the tensors $T_1,\dots,T_n$.) Then it is possible to develop the corresponding "multiplicative" and "additive" theory of eigenvalues and to find estimates from above for the numbers $\max\{k(\mathcal{T}) \mid \mathcal{T} \in \tilde{\Pi}_n^{n_1,\dots,n_r}\}$.

For the basic class $\Pi_n^{n_1,\dots,n_r}$, the corresponding estimate has been found in [K11]; namely, we have

$$k(\mathcal{T}) \leq (m+1)^{(n+1)/2} \quad \text{for each } \mathcal{T} \in \Pi_n^{n_1,\dots,n_r} \qquad (7)$$

where $m = \max\{n_1,\dots,n_r,1\}$.

The last result has an application in the local Riemannian geometry. Let $(M,g)$ be an analytic Riemannian manifold. A point $p \in M$ is said to be free if there is a local isometry $\psi : U_p \longrightarrow U_p'$ such that $\psi(p) = p$ and $p$ is an isolated fixed point of $\psi$. (We speak briefly about a local symmetry $\psi$ at $p$, cf. 0.32.)

It is not difficult to see that, at a free point $p \in M$, a local symmetry $\psi'$ always exists which is periodic. We define the order $\varkappa(p)$ of a free point $p \in M$ as the minimum integer $k \geq 2$ for which a local symmetry $\psi$ at $p$ exists such that $\psi^k = \mathrm{id}$.

If $(M,g)$ is a locally symmetric space, then all points of $M$ are free and $\varkappa(p) = 2$ for all $p \in M$. In the general case, we can give a rough estimate of the number $\varkappa(p)$ from above using our combinatorial results. Put

$$G_p^{(i)} = \{A \in GL(M_p) \mid A(g_p) = g_p,\ A(R_p) = R_p, \dots, A(\nabla^{(i)}R_p) = \nabla^{(i)}R_p\}$$

where $\nabla$ denotes the Riemannian connection in $(M,g)$ and $R$ is the curvature tensor field. Then we have an infinite sequence of Lie groups $G_p^{(o)} \supset G_p^{(1)} \supset \dots$ . Because $G_p^{(o)}$ is compact, the number of distinct groups in our sequence is always finite. By the order of stability $s(p)$ we shall mean the minimum integer $\ell \geq 0$ such that $G_p^{(\ell)} = G_p^{(\ell+1)} = \dots$ etc. Now, we have

Theorem V.33. Let $(M,g)$ be an analytic Riemannian manifold, and $p \in M$ a free point. Then

$$\varkappa(p) \leq (s(p) + 4)^{(1+\dim M)/2}$$

Proof. Because $(M,g)$ is analytic, a non-singular transformation $A: M_p \longrightarrow M_p$ gives rise to a local isometry at $p$ if and only if $A \in \bigcap_{i=0}^{\infty} G_p^{(i)}$. Now, according to the definition of $s(p)$, $\bigcap_{i=0}^{\infty} G_p^{(i)} = G_p^{(s(p))}$. Thus, $A$ coincides with the tangent map $\psi_{*p}$ of a local isometry $\psi$ if and only if $A$ is an automorphism of the tensorial object $(M_p, g_p, R_p, \dots, \nabla^{(s(p))}R_p)$. For the periods of this object we get the same estimate as in Formula (7). In fact, the role of the scalar product $g_p$ is unessential, and the rest is a tensor structure of the type studied above. Here the maximum of the numbers $n_1, \dots, n_r$ is $m = s(p) + 3$, and the result follows. □

Another application is the study of periodic automorphisms of skew-symmetric non-associative real algebras. In the periodic case we get the same numerical estimate as in Theorem V.31. (See [K9] for the details.)

As a rule, our estimates are well-fitting for the "big" subclasses of the basic class $\Pi_n^{n_1,\dots,n_r}$, and they are rather loose for special subclasses of tensor structures (where the tensors $T_1, \dots, T_r$ are subjected to some more complicated relations). Unfortunately, this is also the case for our infinitesimal s-manifolds, or for Lie algebras. In fact, the real growth of the maximum order in Theorems V.31, V.32 may be very far from the exponential one.

-------

<u>Problem V.34.</u> Give an explicit construction of the set $\mathcal{D}^n$ for an arbitrary n.

<u>Problem V.35.</u> As we know from Theorem V.25, all minimal $\theta$-varieties included in $\mathcal{A}^n \cap \mathcal{B}^n$ are finite sets. On the other hand, there are minimal $\theta$-varieties of dimension 1 which are contained in $\mathcal{A}^n \cap (R^n)^o$. (Cf. Chapter VII.) Do also other types of minimal $\theta$-varieties exist in $(C^n)^o$ ?

<u>Problem V.36.</u> Consider the class of all generalized symmetric Riemannian spaces of a given dimension n (without the restriction to the simply connected ones). Are the orders still bounded from above ?

---

References: Primary - [K8],[K9],[K11]
Secondary - [F3],[K6],[Lo2].

# CHAPTER VI

## THE CLASSIFICATION OF GENERALIZED SYMMETRIC RIEMANNIAN SPACES IN LOW DIMENSIONS

### The classification procedure.

Let us describe a procedure for obtaining a complete list of simply connected and irreducible generalized symmetric Riemannian spaces of a given (small) dimension $n$.

First of all, every space in question can be represented by a regular Riemannian s-manifold such that its system of eigenvalues belongs to $\mathfrak{D}^n$ (Theorem V.20), and the latter can be represented by a Riemannian infinitesimal s-manifold (Theorem III.39). Thus we have to classify all Riemannian infinitesimal s-manifolds $(V,g,S,\widetilde{R},\widetilde{T})$ which are irreducible and such that the system of eigenvalues of $S$ belongs to $\mathfrak{D}^n$. According to Theorem IV.43, we can limit ourselves to a subset $\mathfrak{D}^{*n} \subset \mathfrak{D}^n$ consisting of all n-tuples $(\theta_1,\dots,\theta_n) \in \mathfrak{D}^n$ for which the set $\{\theta_1,\dots,\theta_n\}$ is irreducible in the sense of Definition IV.42.

We start our procedure with calculating all the essentially different elements of $\mathfrak{D}^{*n}$. There is a routine and systematic way to perform it:

a) For each $r = 0,1,\dots,[\frac{n}{2}]$ successively, write down the system of equations (5) from V.29.

b) Solve all possible subsystems of $(n-r)$ equations in $(n-r)$ unknows $y^1,\dots,y^{n-r}$ such that each subsystem involves all the equations $2y^i = 1$ for $i = r+1,\dots,n-r$ and has a non-zero determinant.

c) For every $r$, write down all the solutions $[b^1,\dots,b^r,\dots,b^{n-r}]$ such that $0 < b^1 \leq b^2 \leq \dots \leq b^r < \frac{1}{2}$, and form the corresponding n-tuples $(\theta_1,\dots,\theta_n)$.

d) Eliminate all reducible n-tuples $(\theta_i)$.

In practice, the procedure can be shortened by various tricks.

Further, we are interested only in those generalized symmetric spaces which are not locally symmetric. For this reason, we shall seek only infinitesimal s-manifolds $(V,g,S,\widetilde{R},\widetilde{T})$ with $\widetilde{T} \neq 0$ (see Corollary III.42) and we shall also drop the element $(-1,\dots,-1) \in \mathfrak{D}^{*n}$.

Choose a fixed $(\theta_1,\ldots,\theta_n)\in\mathfrak{D}^{*n}$ and a permutation $\wp$ of $\{1,\ldots,n\}$ such that $\wp^2 = \mathrm{id}$, and $\theta_{\wp(i)} = \bar{\theta}_i$. Let $\{U_1,\ldots,U_n\}$ be a basis of $V^c$ such that $U_{\wp(i)} = \bar{U}_i$; define $S\colon V \longrightarrow V$ by the relations $SU_i = \theta_i U_i$, $i=1,\ldots,n$. Let us put $g(U_i,U_j) = \alpha_{ij}$, $\widetilde{T}(U_i,U_j) = \sum_k \beta^k_{ij} U_k$, for $i,j=1,\ldots,n$, where $\alpha_{ij}$, $\beta^k_{ij}$ are arbitrary complex variables, and not all $\beta^k_{ij}$ are allowed to be zero. We express the relations $S(g) = g$, $S(\widetilde{T}) = \widetilde{T}$, $\widetilde{T}(X,Y) = -\widetilde{T}(Y,X)$, $g(X,Y) = g(Y,X)$, and the positivity of $g$, by means of the variables $\alpha_{ij}$, $\beta^k_{ij}$. We also express the property that $g$ and $\widetilde{T}$ are linear extensions of real tensors on $V$. Then we either get a contradiction, or $g$ and $\widetilde{T}$ still depend on a number of variables. Now we try, by a possible change of the basis $\{U_1,\ldots,U_n\}$, to reduce $g$ and $\widetilde{T}$ to a canonical form, minimizing the common number of variables in their expressions. As a rule, we have to distinguish a finite number of types leading to different canonical forms. Indeed, each canonical form defines a family of orbits with respect to the representation of a subgroup $G \subset GL(V)$ in the space of all admissible pairs of tensors $(g,\widetilde{T})$. The specification of all variables in a canonical form means a choice of a particular orbit.

Let us consider a canonical type $(g,\widetilde{T})$ depending on some variables. Let now calculate the Lie algebra $\underline{h}_1$ of all endomorphisms $A$ of $V$ such that $A(S) = A(g) = A(\widetilde{T}) = 0$. It can happen that our canonical type $(g,\widetilde{T})$ splits into a finite number of subtypes with different algebras $\underline{h}_1$. Consider a tensor $\widetilde{R}$, where $\widetilde{R}(U_i,U_j)U_k = \sum_\ell \gamma^\ell_{ijk} U_\ell$ and $\gamma^\ell_{ijk}$ are complex variables. We express first the relations $\widetilde{R}(U_i,U_j) \subset \underline{h}^c_1$ for all $U_i,U_j$ and then all the necessary conditions from III.19 and III.35 - the axioms of a Riemannian infinitesimal s-manifold. We also express the reality conditions for $\widetilde{R}$. We either come to a contradiction, or $\widetilde{R}$ depends on a number of variables. If possible, we minimize this number of variables by a change of the basis $U_1,\ldots,U_n$ (leaving the canonical form of $g$ and $\widetilde{T}$ invariant). Once more, we may have to distinguish several types of tensors $\widetilde{R}$ with different canonical forms.

-------

Having a Riemannian infinitesimal s-manifold $(V,g,S,\widetilde{R},\widetilde{T})$ with $\widetilde{T} \neq 0$, we want to check whether or not the corresponding simply connected g.s. space $(M,g)$ is locally symmetric. For this purpose, we calculate the difference tensor $D$ from $\widetilde{T}$ and $g$ by means of Formula (14), Proposition III.41. Further, we calculate the Riemannian curvature tensor on $V$ using the formula $R(X,Y) = \tilde{R}(X,Y) + [D_X, D_Y] + D_{\tilde{T}(X,Y)}$ for $X, Y \in V$, which is an algebraic analogue of Formula (6), Lemma III.33. Now, $\nabla R \neq 0$ if and only if $D_Z R \neq 0$ for some $Z \in V$.

It remains to carry out the construction of the space $(M,g)$ and of the auxiliary regular s-structure $\{s_x\}$ from the infinitesimal s-manifold $(V,g,S,\tilde{R},\tilde{T})$. In accordance with Theorems I.17 and I.19, we construct first the Lie algebra $\underline{h}$ which is generated by the endomorphisms $\tilde{R}(X,Y)$, $X, Y \in V$, and then the Lie algebra $\underline{g} = V + \underline{h}$, where

$$\left.\begin{aligned} [X,Y] &= (-\tilde{T}(X,Y),\ -\tilde{R}(X,Y)) \\ [A,X] &= AX \\ [A,B] &= AB - BA \end{aligned}\right\} \quad \text{for } X, Y \in V,\ A, B \in \underline{h}.$$

The next step is the geometric realization of the homogeneous space $G/H$. Fortunately, in many cases the Lie algebra $\underline{g}$ is centerless and we can use the adjoint representation. In practice, we often try to find the representation of $\underline{g}$ by infinitesimal affine transformations of a cartesian space $R^k$. Hence we obtain $G$ and $H$ as certain matrix groups. The tensor $g$ gives rise to an invariant metric on $G/H$. Finally, we can always represent the regular s-structure $\{s_x\}$ by an automorphism $\mathfrak{S}$ of the group $G$ (cf. Theorem II.25).

In many cases the homogeneous space $G/H$ has a simple topological structure (e.g. it is diffeomorphic to $R^n$, $S^n$, $R^k \times S^{n-k}$ and similarly). It is always true if $\tilde{R} = 0$ (cf. Chapter II). In these cases we are able to present all the admissible invariant metrics on the underlying manifold $G/H$ in a simple explicit form. Moreover, we can give a "typical" symmetry at the initial point of $(G/H,g)$ in the form of a linear transformation.

-------

It is not necessary to apply our procedure in dimension $n = 2$, where we have the following result:

<u>Theorem VI.1</u>. Any generalized symmetric Riemannian space of dimension $n = 2$ (simply connected or not) is Riemannian symmetric.

Proof. $(M,g)$ is homogeneous and of constant curvature. Hence it is homothetic to one of the following spaces: the Euclidean plane $E^2$, the cylinder $S^1 \times E^1$, the flat torus $T^2$, the sphere $S^2$, the projective plane $P^2 = S^2/\{\pm I\}$, or the hyperbolic plane $H^2$. (See [Wo], Chapter 2.)

-------

In the following paragraphs we shall solve the classification problem in the dimensions $n = 3, 4, 5$. Here the complete proofs will be given only for $n = 3, 4$; for dimension 5 we shall draw a short outline of the procedure. We refer to [K3] for the details. A short remark touches the classification for $n = 6$.

For the sake of brevity, the generalized symmetric Riemannian spaces which are not Riemannian symmetric in the classical sense will be said to be proper.

## Dimension $n = 3$.

Theorem VI.2. Any proper simply connected generalized symmetric Riemannian space $(M,g)$ of dimension $n = 3$ is of order $k = 4$. It is irreducible and described as follows:
The underlying homogeneous space is the matrix group

$$\left\| \begin{matrix} e^{-z} & 0 & x \\ 0 & e^{z} & y \\ 0 & 0 & 1 \end{matrix} \right\| .$$

$(M,g)$ is the space $R^3[x,y,z]$ with the Riemannian metric $g = e^{2z}dx^2 + e^{-2z}dy^2 + \lambda^2 dz^2$, where $\lambda > 0$ is a constant. A typical symmetry of order 4 at the origin $[0,0,0]$ is the transformation $x' = -y$, $y' = x$, $z' = -z$.

Proof. The only essentially different elements in $\mathfrak{D}^{*3} = \mathfrak{D}^3$ are the triplets $(-1, -1, -1)$ and $(\sqrt{-1}, -\sqrt{-1}, -1)$. Here the first triplet yields the classical symmetric spaces. We shall put $i = \sqrt{-1}$ in this Chapter.

Let $V$ be a 3-dimensional real vector space and $S: V \longrightarrow V$ a linear transformation with the eigenvalues $i, -i, -1$; $g$ an inner product on $V$ such that $S(g) = g$, and $\tilde{T} \neq 0$ a tensor of type $(1,2)$ such that $\tilde{T}(Y,X) = -\tilde{T}(X,Y)$ and $S(\tilde{T}) = \tilde{T}$. We shall denote by the same symbols the linear extensions of $S$, $g$ and $\widetilde{T}$ to the complexification $V^c = V\otimes_R C$ of $V$.

Let $U \in V^c$ be a complex eigenvector such that $SU = iU$ and $W \in V$ a real eigenvector such that $SW = -W$. Then $S\bar{U} = -i\bar{U}$. The condition $S(g) = g$ means that $g(SZ,SZ') = g(Z,Z')$ for every $Z, Z' \in V^c$ and hence $g(U,U) = g(\bar{U},\bar{U}) = g(W,U) = g(W,\bar{U}) = 0$, $g(U,\bar{U}) = a^2 > 0$, $g(W,W) = b^2 > 0$, where $a$, $b$ are real variables. Further, the property $S(\widetilde{T}) = \widetilde{T}$ means that $\widetilde{T}(SZ,SZ') = S(\widetilde{T}(Z,Z'))$ for every $Z,Z' \in V^c$. Hence it follows $\widetilde{T}(U,\bar{U}) = 0$, $\widetilde{T}(U,W) = \alpha\bar{U}$, $\widetilde{T}(\bar{U},W) = \bar{\alpha}U$, where $\alpha \neq 0$ is a complex variable.

Suppose $\alpha = \rho e^{2i\varphi}$ with $\rho > 0$ and put $U' = (1/a)e^{-i\varphi}U$, $W' = (1/\rho)W$. Then replacing $U$ and $W$ by $U'$ and $W'$ we can reduce the variables $a^2$ and $\alpha$ to 1. We conclude that, for each triplet $(S,g,\widetilde{T})$ of tensors with the required properties on $V$, there is a basis $(U,\bar{U},W)$ of $V^c$ $(W \in V)$ such that

$$\begin{aligned} &SU = iU, \quad S\bar{U} = -i\bar{U}, \quad SW = -W \\ &g(U,U) = g(\bar{U},\bar{U}) = g(W,U) = g(W,\bar{U}) = 0, \\ &g(U,\bar{U}) = 1, \quad g(W,W) = \lambda^2 > 0, \\ &\widetilde{T}(U,\bar{U}) = 0, \quad \widetilde{T}(U,W) = \bar{U}, \quad \widetilde{T}(\bar{U},W) = U . \end{aligned} \tag{1}$$

We have obtained the canonical form of an arbitrary admissible triplet $(S,g,\widetilde{T})$, $\widetilde{T} \neq 0$, and we can see easily that $\lambda > 0$ is an invariant of this canonical form. It means that two triplets $(S_j,g_j,\widetilde{T}_j)$ $(j = 1,2)$ with different $\lambda$'s cannot overlap by means of a linear transformation $f\colon V \longrightarrow V$.

Now, let $\underline{k}$ denote the Lie algebra of all real endomorphisms $A\colon V^c \longrightarrow V^c$ which, as derivations, satisfy $A(S) = A(g) = A(\widetilde{T}) = 0$. The relation $A(S) = 0$ means $A\cdot S = S\cdot A$, and thus $AU = uU$, $A\bar{U} = \bar{u}\bar{U}$, $AW = wW$ ($w$ real). Further, the relation $A(g) = 0$ means that $g(AZ,Z') + g(Z,AZ') = 0$ for every $Z, Z' \in V^c$ and hence $u + \bar{u} = 0$, $w = 0$. Finally, $A(\widetilde{T}) = 0$ implies $A(\widetilde{T}(U,W)) = \widetilde{T}(AU,W) + \widetilde{T}(U,AW)$ and we get $u = \bar{u}$. Consequently, $u = w = 0$ and $\underline{k} = (0)$.

If $(V,g,S,\widetilde{R},\widetilde{T})$ is an infinitesimal s-manifold, $\widetilde{T} \neq 0$, we can see from the conditions $\widetilde{R}(X,Y)(S) = \widetilde{R}(X,Y)(g) = \widetilde{R}(X,Y)(\widetilde{T}) = 0$ that $\widetilde{R}(X,Y) \in \underline{k}$ for every $X, Y \in V$ and thus $\widetilde{R} = 0$. Conversely, each collection $(V,g,S,0,\widetilde{T})$ with tensors $S$, $g$, $\widetilde{T}$ given by (1) is an infinitesimal s-manifold. Thus all infinitesimal s-manifolds on $V$ with $\widetilde{T} \neq 0$ are of the above form and they depend on a single real parameter $\lambda > 0$.

We can calculate easily that $(D_U R)(U,\bar{U}) \neq 0$ and thus the corresponding generalized symmetric Riemannian spaces $(M,g)$ are not locally symmetric, i.e., they are proper. Further, they are all irreducible: suppose that one of our spaces $(M,g)$ is not irreducible. Then according to the de Rham decomposition theorem we should have $(M,g) = (M_1,g_1)\times(M_2,g_2)$, where $(M_i,g_i)$ are generalized symmetric of dimensions $\leq 2$, hence Riemannian symmetric, and $(M,g)$ should be also Riemannian symmetric - a contradiction.

Let us construct the Lie algebras $\underline{h}$ and $\underline{g}$. Obviously $\underline{h} = (0)$ because $\tilde{R} = 0$, and the bracket in $\underline{g}$ is given by the formula $[X_1,X_2] = -\tilde{T}(X_1,X_2)$ for every $X_1, X_2 \in V$. Put $U = (X + iY)/\sqrt{2}$, $W = Z$ with $X, Y, Z \in V$. Then $g(X,X) = g(Y,Y) = 1$, $g(Z,Z) = \lambda^2$, and $X, Y, Z$ are mutually orthogonal. The transformation $S$ satisfies $SX = -Y$, $SY = X$, $SZ = -Z$. Further, we obtain the multiplication table for $\underline{g}$ in the form

$$[X,Y] = 0, \quad [X,Z] = X, \quad [Y,Z] = -Y.$$

We can find easily a representation of $\underline{g}$ by infinitesimal transformations of $R^2[x,y]$, namely $X = \partial/\partial x$, $Y = \partial/\partial y$, $Z = x(\partial/\partial x) - y(\partial/\partial y)$. Hence we get a finite representation, namely the group of "hyperbolic motions"

$$G = \begin{Vmatrix} e^{-c} & 0 & a \\ 0 & e^{c} & b \\ 0 & 0 & 1 \end{Vmatrix} .$$

The underlying manifold of the group $G$ is the Cartesian space $R^3[a,b,c]$. The elements $X, Y, Z \in \underline{g}$ can be now represented by the left-invariant vector fields $e^{-c}(\partial/\partial a)$, $e^{c}(\partial/\partial b)$, $\partial/\partial c$ on $G$ respectively. Then the inner product $g$ on $V = \underline{g}$ induces an invariant Riemannian metric on $R^3[a,b,c]$:

$$g = e^{2c}da^2 + e^{-2c}db^2 + \lambda^2 dc^2, \qquad \lambda > 0.$$

Finally, a symmetry $S_o$ of order 4 at the origin $[0,0,0] \in R^3$ is given by $a' = -b$, $b' = a$, $c' = -c$. (Cf. also Chapter 0.) □

## D i m e n s i o n  n = 4 .

Theorem VI.3. Any proper, simply connected and irreducible generalized symmetric Riemannian space $(M,g)$ of dimension $n = 4$

is of order $k = 3$ and it is described as follows:
The underlying homogeneous space is of the form

$$\begin{Vmatrix} a & b & u \\ c & d & v \\ 0 & 0 & 1 \end{Vmatrix} \Big/ \begin{Vmatrix} \cos t & -\sin t & 0 \\ \sin t & \cos t & 0 \\ 0 & 0 & 1 \end{Vmatrix} \quad \text{where} \quad \det \begin{Vmatrix} a & b \\ c & d \end{Vmatrix} = 1.$$

$(M,g)$ is the space $R^4[x,y,u,v]$ with the Riemannian metric

$$g = (-x + \sqrt{(x^2 + y^2 + 1)})du^2 + (x + \sqrt{(x^2 + y^2 + 1)})dv^2 - 2ydudv +$$
$$+ \lambda^2 \left[ \frac{(1 + y^2)dx^2 + (1 + x^2)dy - 2xydxdy}{1 + x^2 + y^2} \right] \qquad (\lambda > 0).$$

A typical symmetry of order 3 at the origin $[0,0,0,0]$ is the transformation

$$u' = \cos\frac{2\pi}{3}\cdot u - \sin\frac{2\pi}{3}\cdot v, \qquad v' = \sin\frac{4\pi}{3}\cdot u + \cos\frac{4\pi}{3}\cdot v$$
$$x' = \cos\frac{2\pi}{3}\cdot x - \sin\frac{2\pi}{3}\cdot y, \qquad y' = \sin\frac{4\pi}{3}\cdot x + \cos\frac{4\pi}{3}\cdot y.$$

Proof. The following assertion is left to the reader as an exercise:

Proposition VI.4. The essentially different elements of $\mathfrak{D}^{*4} = \mathfrak{D}^4$ are the following:

a) $(\theta,\theta,\theta^2,\theta^2)$, $\quad \theta = e^{2\pi i/3}$

b) $(\theta,\theta^2,\theta^3,\theta^4)$, $\quad \theta = e^{2\pi i/5}$

c) $(i, -i, -1, -1)$

d) $(-1, -1, -1, -1)$.

We shall examine our systems of eigenvalues separately.

(A) The system of eigenvalues $(\theta, \theta, \bar\theta = \theta^2, \bar\theta = \theta^2)$, $\quad \theta = e^{2\pi i/3}$.

Let $V$ be a 4-dimensional vector space, $V^c$ its complexification and $S: V^c \longrightarrow V^c$ a real linear transformation with the eigenvalues given above. Further, let $g$, $\tilde T$ be tensors on $V$ satisfying usual conditions; denote their linear extensions to $V^c$ by the same letters. We can find a basis $(U_1,U_2,\bar U_1,\bar U_2)$ of eigenvectors in $V^c$ such that $SU_1 = \theta U_1$, $SU_2 = \theta U_2$, $S\bar U_1 = \bar\theta \bar U_1$, $S\bar U_2 = \bar\theta \bar U_2$, $g(U_1,\bar U_1) = g(U_2,\bar U_2) = 1$, $g(U_1,\bar U_2) = 0$. Further, owing to $S(g) = g$ we have $g(U_j,U_k) = g(\bar U_j,\bar U_k) = 0$. For the tensor $\tilde T$ we get $\tilde T(U_1,U_2) = \alpha\bar U_1 + \beta\bar U_2$, $\tilde T(\bar U_1,\bar U_2) = \bar\alpha U_1 + \bar\beta U_2$, $\tilde T(U_j,\bar U_k) = 0$, owing to the relation $S(\tilde T) = \tilde T$. Here $\alpha$, $\beta$ are complex variables.

If $\widetilde{T} \neq 0$, then $\rho^2 = \alpha\bar{\alpha} + \beta\bar{\beta} > 0$. Let us replace $U_1, U_2$ by the new eigenvectors $U_1' = (\bar{\alpha}U_1 + \bar{\beta}U_2)/\rho$, $U_2' = (-\beta U_1 + \alpha U_2)/\rho^2$ and then write again $U_1$, $U_2$ instead of $U_1'$, $U_2'$. We get the following:

$$\left.\begin{aligned} &SU_1 = \theta U_1, \quad SU_2 = \theta U_2, \quad S\bar{U}_1 = \bar{\theta}\bar{U}_1, \quad S\bar{U}_2 = \bar{\theta}\bar{U}_2, \\ &g(U_1,\bar{U}_1) = 1, \quad g(U_2,\bar{U}_2) = 1/\rho^2, \quad g(U_1,\bar{U}_2) = g(\bar{U}_1,U_2) = 0, \\ &g(U_j,U_k) = 0, \\ &\widetilde{T}(U_1,U_2) = \bar{U}_1, \quad \widetilde{T}(\bar{U}_1,\bar{U}_2) = U_1, \quad \widetilde{T}(U_j,\bar{U}_k) = 0. \end{aligned}\right\} \quad (2)$$

We have obtained a canonical form of the triplet $(g,S,\widetilde{T})$, $\widetilde{T} \neq 0$; all these triplets depend on a real invariant $\rho$.

Let $\underline{k}$ be the Lie algebra of all real endomorphisms $A$ of $V^c$ which as derivations annihilate $S$, $g$, and $\widetilde{T}$.

$A(S) = 0$ means that $AU_j = \sum a_j^k U_k$, $A\bar{U}_j = \sum \bar{a}_j^k \bar{U}_k$ $(j, k = 1,2)$,

$A(g) = 0$ means that $a_1^1 + \bar{a}_1^1 = 0$, $a_2^2 + \bar{a}_2^2 = 0$, $a_1^2 + \rho^2\bar{a}_2^1 = 0$.

$A(\widetilde{T}) = 0$ implies, in particular, $A(\widetilde{T}(U_1,U_2) = \widetilde{T}(AU_1,U_2) + \widetilde{T}(U_1,AU_2)$, whence $a_1^1 + a_2^2 = \bar{a}_1^1$, $\bar{a}_1^2 = 0$. We get $a_1^2 = a_2^1 = 0$, $a_2^2 = -2a_1^1$, $a_j^j + \bar{a}_j^j = 0$. Consequently, the Lie algebra $\underline{k}$ is 1-dimensional and generated by the endomorphism $B$ satisfying

$$BU_1 = iU_1, \quad BU_2 = -2iU_2, \quad B\bar{U}_1 = -i\bar{U}_1, \quad BU_2 = 2i\bar{U}_2 \quad (3).$$

Let $(V,g,S,\widetilde{R},\widetilde{T})$ be an infinitesimal s-manifold with the tensors $g$, $S$, $\widetilde{T}$ satisfying (2). For every $X, Y \in V$ we have $\widetilde{R}(X,Y) \in \underline{k}$, and for every $Z, Z' \in V^c$ it follows $\widetilde{R}(Z,Z') \in \underline{k} \otimes_R C$. The first Bianchi identity $\mathfrak{S}(\widetilde{R}(Z,Z')Z'') = \mathfrak{S}(\widetilde{T}(\widetilde{T}(Z,Z'),Z''))$ must hold in $V^c$ and, in particular, we get

$$\widetilde{R}(U_1,U_2)\bar{U}_1 + \widetilde{R}(U_2,\bar{U}_1)U_1 + \widetilde{R}(\bar{U}_1,U_1)U_2 = 0,$$
$$\widetilde{R}(U_1,U_2)\bar{U}_2 + \widetilde{R}(U_2,\bar{U}_2)U_1 + \widetilde{R}(\bar{U}_2,U_1)U_2 = U_1.$$

Hence

$$\left.\begin{aligned} &\widetilde{R}(U_1,U_2) = \widetilde{R}(\bar{U}_1,\bar{U}_2) = \widetilde{R}(\bar{U}_1,U_2) = \widetilde{R}(U_1,\bar{U}_2) = \widetilde{R}(U_1,\bar{U}_1) = 0, \\ &\widetilde{R}(U_2,\bar{U}_2) = -iB. \end{aligned}\right\} \quad (4)$$

We check easily that $B(\widetilde{R}) = 0$ and, in particular, $\widetilde{R}(Z,Z')(\widetilde{R}) = 0$ for every $Z, Z' \in V^c$. The second Bianchi identity $\mathfrak{S}(\widetilde{R}(T(Z,Z'),Z'') = 0$ also holds. Consequently, the collection $(V,g,S,\widetilde{R},\widetilde{T})$ given by (2) and (4) is an infinitesimal s-manifold for each $\rho > 0$. At the same time we see that the Lie algebra $\underline{h}$ generated by $\tilde{R}(X,Y)$ coincides with $\underline{k}$.

Calculating $D$ and $R$ we check that $(D_{U_1}R)(U_1,U_2)\bar{U}_1 \neq 0$, and thus the corresponding generalized symmetric Riemannian spaces are not locally symmetric. Thus they are proper and of order 3.

Let us determine an orthogonal basis $(X_1,Y_1,X_2,Y_2)$ of $V$ by $U_1 = (X_1 + iY_1)/\sqrt{2}$, $U_2 = (X_2 + iY_2)/2$. Then

$$g(X_1,X_1) = g(Y_1,Y_1) = 1, \quad g(X_2,X_2) = g(Y_2,Y_2) = 2/\rho^2 \qquad (5).$$

The Lie algebra $\underline{g} = V + \underline{h}$ obeys the rule

$$[X,Y] = (-\tilde{T}(X,Y),\ -\tilde{R}(X,Y)) \quad \text{for } X, Y \in V$$

$$[X,A] = -AX \quad \text{for } X \in V,\ A \in \underline{h}.$$

With respect to the basis $\{X_1,Y_1,X_2,Y_2,B\}$ we obtain without difficulties the following table of multiplication:

$$[X_1,Y_1] = 0, \quad [X_1,X_2] = -X_1, \quad [X_1,Y_2] = Y_1, \quad [Y_1,Y_2] = Y_1,$$
$$[Y_1,Y_2] = X_1, \quad [X_2,Y_2] = -2B, \quad [X_1,B] = Y_1, \quad [Y_1,B] = -X_1,$$
$$[X_2,B] = -2Y_2, \quad [Y_2,B] = 2X_2.$$

We shall try to represent the vectors of $\underline{g}$ by infinitesimal affine transformations of the plane $R^2[x,y]$. Putting $X_1 = \partial/\partial x$, $Y_1 = \partial/\partial y$ we calculate easily that $X_2 = y\frac{\partial}{\partial y} - x\frac{\partial}{\partial x}$, $Y_2 = x\frac{\partial}{\partial y} + y\frac{\partial}{\partial x}$, $B = x\frac{\partial}{\partial y} - y\frac{\partial}{\partial x}$. The corresponding Lie group is the group of (positive) equiaffine transformations of the plane, or equivalently, the group $G$ of all matrices of the form $\begin{Vmatrix} a & b & \alpha \\ c & d & \beta \\ 0 & 0 & 1 \end{Vmatrix}$ satisfying the relation $ad - bc = 1$. Our basis of the Lie algebra $\underline{g}$ is represented by the following left-invariant vector fields on $G$:

$$X_1 = a\frac{\partial}{\partial\alpha} + c\frac{\partial}{\partial\beta}, \quad Y_1 = b\frac{\partial}{\partial\alpha} + d\frac{\partial}{\partial\beta}, \quad X_2 = a\frac{\partial}{\partial a} - b\frac{\partial}{\partial b} + c\frac{\partial}{\partial c} - d\frac{\partial}{\partial d},$$
$$Y_2 = -(a\frac{\partial}{\partial b} + b\frac{\partial}{\partial a} + c\frac{\partial}{\partial d} + d\frac{\partial}{\partial c}), \quad B = a\frac{\partial}{\partial b} - b\frac{\partial}{\partial a} + c\frac{\partial}{\partial d} - d\frac{\partial}{\partial c}.$$

The subgroup $H$ of $G$ generated by the infinitesimal transformation $B$ is the group of all matrices of the form $\begin{Vmatrix} \cos t & -\sin t & 0 \\ \sin t & \cos t & 0 \\ 0 & 0 & 1 \end{Vmatrix}$.

Consider the mapping $\pi$ of $G$ into the cartesian space $R^4[u,v,x,y]$ given by $u=\alpha$, $v=\beta$, $x=(a^2+b^2-c^2-d^2)/2$, $y=ac+bd$. We see easily that the map $\pi$ is the projection map of a principal fibre bundle with the base space $R^4$ and structural group $H$. In particular, the homogeneous space $G/H$ is diffeomorphic to $R^4$. For any $m\in G$ the projections of the tangent vectors $X_1$, $Y_1$, $X_2$, $Y_2$ into $G/H = R^4$ are the tangent vectors

$$\tilde{X}_1 = a\frac{\partial}{\partial u} + c\frac{\partial}{\partial v}, \quad \tilde{Y}_1 = b\frac{\partial}{\partial u} + d\frac{\partial}{\partial v},$$

$$\tilde{X}_2 = (a^2+d^2-b^2-c^2)\frac{\partial}{\partial x} + 2(ac-bd)\frac{\partial}{\partial y},$$

$$\tilde{Y}_2 = 2(cd-ab)\frac{\partial}{\partial x} - 2(ad+bc)\frac{\partial}{\partial y}$$

at the point $\pi(m)$. These tangent vectors vary with $m$.

On the other hand, the corresponding metric $g$ on $G/H$ is $G$-invariant. This means that the values $g(\tilde{X}_i,\tilde{X}_j)$, $g(\tilde{Y}_i,\tilde{Y}_j)$, $g(\tilde{X}_i,\tilde{Y}_j)$ are independent of the choice of $m\in G$, namely we have $g(\tilde{X}_1,\tilde{X}_1) = g(\tilde{Y}_1,\tilde{Y}_1) = 1$, $g(\tilde{X}_2,\tilde{X}_2) = g(\tilde{Y}_2,\tilde{Y}_2) = 2/\rho^2$, $g(\tilde{X}_1,\tilde{X}_2) = g(\tilde{Y}_1,\tilde{Y}_2) = g(\tilde{X}_1,\tilde{Y}_2) = g(\tilde{X}_2,\tilde{Y}_1) = 0$ (cf. Formula (5)). Expressing $\partial/\partial u$, $\partial/\partial v$ by means of $\tilde{X}_1$, $\tilde{Y}_1$ and $\partial/\partial x$, $\partial/\partial y$ by means of $\tilde{X}_2$, $\tilde{Y}_2$ we obtain by an easy calculation the same invariant metric as in the statement of our Theorem.

Further, the action of $G$ on $R^4[u,v,x,y]$ is induced by the left translations of $G$ onto itself. Hence the subgroup $H$ acts on $R^4$ as a group of isometries according to the rule

$$u' = \cos\varphi\cdot u - \sin\varphi\cdot v, \qquad v' = \sin\varphi\cdot u + \cos\varphi\cdot v$$

$$x' = \cos 2\varphi\cdot x - \sin 2\varphi\cdot y, \qquad y' = \sin 2\varphi\cdot x + \cos 2\varphi\cdot y .$$

Now, $H$ is the isotropy subgroup of $G$ at the origin, and according to Lemma 0.14, all transformations $\Phi\in H$ with $\varphi\neq n\pi$ are symmetries at the origin, and they define regular s-structures on $(R^4,g)$. In particular, our space is k-symmetric for every $k\geq 3$, and the last assertion of Theorem VI.3 follows.

Finally, we need to check that our Riemannian spaces are irreducible. Let $(M,g)$ be a simply connected space belonging to our family and suppose $(M,g) = (M_1,g_1)\times(M_2,g_2)$. Here $M_1$, $M_2$ cannot be of dimension 2, otherwise they would be locally symmetric and $(M,g)$ would be locally symmetric, too. Suppose now that $\dim M_1 = 3$, $\dim M_2 = 1$, $M_1$ being proper. Then $(M_1,g_1)$ is one of the spaces from Theorem VI.2. As we know from Chapter 0, $(M_1,g_1)$ does not possess infinitesimal rotations, and hence $\dim I^\bullet(M_1,g_1) = 3$.

Further, $(M_2,g_2)$ is isometric to the euclidean line and hence $\dim I^\bullet(M_2,g_2) = 1$. A well-known theorem says that $I^\bullet(M) = I^\bullet(M_1) \times I^\bullet(M_2)$ (see [KN II]) and hence $\dim I^\bullet(M,g) = 4$, a contradiction, because $I^\bullet(M,g) \supset G$.

(B) The system of eigenvalues $(\theta,\ \bar{\theta}=\theta^4,\ \tau=\theta^3,\ \bar{\tau}=\theta^2)$, $\theta = e^{2\pi i/5}$.

Let $(V,S,g,\tilde{R},\tilde{T})$, $\tilde{T} \neq 0$, be an infinitesimal s-manifold with the above system of eigenvalues and let $U_1$, $U_2$ be complex unit eigenvectors corresponding to $\theta, \tau$ respectively. Then

$$SU_1 = \theta U_1, \quad SU_2 = \theta^3 U_2, \quad S\bar{U}_1 = \theta^4 \bar{U}_1, \quad S\bar{U}_2 = \theta^2 \bar{U}_2,$$

$$g(U_1,\bar{U}_1) = g(U_2,\bar{U}_2) = 1, \quad g(U_1,U_2) = g(U_1,\bar{U}_2) = 0;$$

and finally, $S(\tilde{T}) = \tilde{T}$ implies

$$\tilde{T}(U_1,U_2) = \alpha\bar{U}_1, \quad \tilde{T}(U_1,\bar{U}_2) = \beta U_2, \quad \tilde{T}(\bar{U}_1,\bar{U}_2) = \bar{\alpha}U_1, \quad \tilde{T}(\bar{U}_1,U_2) = \bar{\beta}U_2,$$

where $\alpha, \beta$ are complex numbers, not both equal to zero.

If $\alpha \neq 0$, $\beta = 0$, then we can reduce the couple $(g,\tilde{T})$ to the canonical form (2), and we derive also (3) and (4). We obtain the same spaces as those sub A), equipped with symmetries of order 5. If $\alpha = 0$, $\beta \neq 0$, we introduce a new basis $\{U_1', U_2'\} = \{U_2, \bar{U}_1\}$, we get the new parameters $\alpha' = -\bar{\beta}$, $\beta' = -\alpha$, and all is reduced to the first case.

Let us now suppose that $\alpha\beta \neq 0$, and calculate the Lie algebra $\underline{k}$ of all real endomorphisms $A$ of $V^c$ such that $A(g) = A(S) = A(\tilde{T}) = 0$.

$A(S) = 0$ means that $AU_1 = \lambda U_1$, $A\bar{U}_1 = \bar{\lambda}\bar{U}_1$, $AU_2 = \mu U_2$, $AU_2 = \bar{\mu}\bar{U}_2$,

$A(g) = 0$ implies $\lambda + \bar{\lambda} = 0$, $\mu + \bar{\mu} = 0$,

$A(\tilde{T}) = 0$ implies $\alpha(\bar{\lambda} - \lambda - \mu) = 0$, $\beta(\mu - \lambda - \bar{\mu}) = 0$.

Hence $\lambda = \mu = 0$ and $\underline{k} = (0)$. This means that $\tilde{R} = 0$, and the first Bianchi identity $\mathfrak{S}(\tilde{T}(\tilde{T}(Z,Z'),Z'')) = 0$ holds. On the other hand,

$$\mathfrak{S}(\tilde{T}(\tilde{T}(U_1,U_2),\bar{U}_1) = \beta\bar{\beta}U_2, \qquad \mathfrak{S}(\tilde{T}(\tilde{T}(U_1,U_2),\bar{U}_2) = \alpha\bar{\alpha}U_1$$

and hence $\alpha = \beta = 0$, a contradiction. We cannot obtain new infinitesimal s-manifolds in this way.

(C) The system of eigenvalues (i, -i, -1, -1).

Let $(V,g,S,\tilde{R},\tilde{T})$, $\tilde{T} \neq 0$, be an infinitesimal s-manifold with the eigenvalues given above. Choose complex eigenvectors $U$, $\bar{U} \in V^c$ corresponding to $i$, $-i$ respectively such that $g(U,\bar{U}) = 1$. Let $H$ be the (real) eigenspace in $V$ corresponding to $-1$, and let $V_1, V_2 \in H$ be such that $g(V_1,V_1) = g(V_2,V_2) = 1$, $g(V_1,V_2) = 0$. From the relation $S(T(Z,Z')) = T(SZ,SZ')$ we get

$$\tilde{T}(U,\bar{U}) = 0, \quad \tilde{T}(V_1,V_2) = 0, \quad \tilde{T}(U,V_1) = \alpha\bar{U}, \quad \tilde{T}(U,V_2) = \beta\bar{U} .$$

Here $\alpha, \beta$ are complex numbers, not both equal to zero.

Now, calculating the Lie algebra $\underline{k}$ given by $A(S) = A(\tilde{T}) = A(g) = 0$ we come to the following conclusions: if $\beta \pm i\alpha \neq 0$, then our infinitesimal s-manifold is reducible, and $\tilde{R} = 0$. If $\beta = \pm i\alpha$ we obtain the same generalized symmetric space as in case A).

(D) The system of eigenvalues $(-1, -1, -1, -1)$.

This system yields only Riemannian symmetric spaces and therefore we shall neglect it.

## D i m e n s i o n n = 5 .

Theorem VI.5. All proper, simply connected and irreducible generalized symmetric Riemannian spaces $(M,g)$ of dimension $n = 5$ are of order $k = 4$ or $k = 6$, and of the following 12 types:

Type 1.

The underlying homogeneous space is the matrix group $\begin{Vmatrix} 1 & 0 & 0 & x \\ 0 & 1 & 0 & y \\ u & v & 1 & z \\ 0 & 0 & 0 & 1 \end{Vmatrix}$.

$(M,g)$ is the space $R^5(x,y,z,u,v)$ with the Riemann metric

$$g = dx^2 + dy^2 + du^2 + dv^2 + \lambda^2(x\,du - y\,dv + dz)^2 \qquad (\lambda > 0).$$

The typical symmetry at the point $(0,\ldots,0)$ is the transformation $x' = -y$, $y' = x$, $z' = -z$, $u' = -v$, $v' = u$.

Type 2.

The underlying homogeneous space is a matrix group depending on two real parameters $\lambda_1 > 0$, $\lambda_2 \geq 0$ :

$$\begin{Vmatrix} e^{\lambda_1 t} & 0 & 0 & 0 & x \\ 0 & e^{-\lambda_1 t} & 0 & 0 & y \\ 0 & 0 & e^{\lambda_2 t} & 0 & z \\ 0 & 0 & 0 & e^{-\lambda_2 t} & w \\ 0 & 0 & 0 & 0 & 1 \end{Vmatrix} .$$

$(M,g)$ is the space $R^5(x,y,z,w,t)$ with the Riemann metric

$$g = e^{-2\lambda_1 t}dx^2 + e^{2\lambda_1 t}dy^2 + e^{-2\lambda_2 t}dz^2 + e^{2\lambda_2 t}dw^2 + dt^2 +$$
$$+ 2\alpha\left[e^{-(\lambda_1+\lambda_2)t}dxdz + e^{(\lambda_1+\lambda_2)t}dydw\right] +$$
$$+ 2\beta\left[e^{(\lambda_1-\lambda_2)t}dydz - e^{(\lambda_2-\lambda_1)t}dxdw\right].$$

Here either $\lambda_1 > \lambda_2 > 0, \quad \alpha \geq 0, \quad \beta \geq 0, \quad \alpha^2 + \beta^2 < 1,$

or $\lambda_1 = \lambda_2 > 0, \quad \alpha = 0, \quad 0 \leq \beta < 1,$

or $\lambda_1 > 0, \quad \lambda_2 = 0, \quad \alpha = 0, \quad 0 < \beta < 1.$

The typical symmetry at the point $(0,\ldots,0)$ is the transformation $x' = -y, \quad y' = x, \quad z' = -w, \quad w' = z, \quad t' = -t.$

Type 3.

The underlying homogeneous space $M$ is $SO(3,C)/SO(2)$, where $SO(3,C)$ denotes the special complex orthogonal group and $SO(2)$ denotes the subgroup $\left\|\begin{array}{c:c} SO(2) & 0 \\ \hdashline 0 & 1 \end{array}\right\|$ of $SO(3,C)$.

The Riemann metric $g$ in $M$ is induced by the following real invariant positive semi-definite form on the group $GL(3,C)$ of all regular complex matrices $\left\|\begin{matrix} a_1 & a_2 & a_3 \\ b_1 & b_2 & b_3 \\ c_1 & c_2 & c_3 \end{matrix}\right\|$ :

$$\tilde{g} = \lambda^2(\omega_1\bar{\omega}_1 + \omega_2\bar{\omega}_2) + \gamma((\omega_1)^2 + (\bar{\omega}_1)^2 + (\omega_2)^2 + (\bar{\omega}_2)^2) + \mu^2\left(\frac{\omega_3 - \bar{\omega}_3}{i}\right)^2$$

where $\omega_1 = a_2da_3 + b_2db_3 + c_2dc_3, \qquad \omega_2 = a_3da_1 + b_3db_1 + c_3dc_1$ ,

$\omega_3 = a_1da_2 + b_1db_2 + c_1dc_2$, and $\lambda, \gamma, \mu$ are real parameters satisfying $\lambda > 0, \ \mu > 0, \ |2\gamma| < \lambda^2$.

The typical symmetry at the origin of $M$ is induced by the following transformation of $GL(3,C)$: $\left\|\begin{matrix} a_1 & a_2 & a_3 \\ b_1 & b_2 & b_3 \\ c_1 & c_2 & c_3 \end{matrix}\right\| \longmapsto \left\|\begin{matrix} \bar{b}_2 & -\bar{b}_1 & \bar{b}_3 \\ -\bar{a}_2 & \bar{a}_1 & -\bar{a}_3 \\ \bar{c}_2 & -\bar{c}_1 & \bar{c}_3 \end{matrix}\right\|$

Type 4.

The underlying homogeneous space is a complex matrix group depending on a complex parameter : $\left\|\begin{matrix} e^t & 0 & z \\ 0 & e^{-t} & w \\ 0 & 0 & 1 \end{matrix}\right\|$ Here $z$, $w$ denote complex variables and $t$ a real variable.

$(M,g)$ is the space $C^2(z,w)\times R^1(t)$ with a (real) Riemann metric

$$g = e^{-(\lambda+\bar\lambda)t}dz d\bar z + e^{(\lambda+\bar\lambda)t}dw d\bar w + (dt)^2 + 2\left[e^{(\bar\lambda-\lambda)t}dz d\bar w + e^{(\lambda-\bar\lambda)t}d\bar z dw\right] +$$
$$+\alpha e^{-2\lambda t}(dz)^2 + \bar\alpha e^{-2\bar\lambda t}(d\bar z)^2 - \alpha e^{2\lambda t}(dw)^2 - \bar\alpha e^{2\bar\lambda t}(d\bar w)^2.$$

Here $\alpha$ is another complex parameter, $\mu \geq 0$ a real parameter and $\alpha\bar\alpha + \mu^2 < \frac{1}{4}$. In case that $\lambda + \bar\lambda = 0$ we have $\alpha = 0$ and $\mu \neq 0$. The typical symmetry at the point $(0,0,0)$ is the transformation $z' = iw$, $w' = iz$, $t' = -t$.

Types 5a, 5b.

The underlying homogeneous space $M$ is either $\dfrac{SO(3)\times SO(3)}{SO(2)}$ or $\dfrac{SO(1,2)\times SO(1,2)}{SO(2)}$, where $SO(2)$ denotes the subgroup of all matrix

pairs of the form $\left(\begin{Vmatrix}\cos t & -\sin t & 0\\ \sin t & \cos t & 0\\ 0 & 0 & 1\end{Vmatrix}, \begin{Vmatrix}\cos t & \sin t & 0\\ -\sin t & \cos t & 0\\ 0 & 0 & 1\end{Vmatrix}\right)$.

The Riemann metric $g$ is induced by the following real invariant positive semi-definite form on the group $GL(3,R)\times GL(3,R)$ of all

regular matrix pairs $\left(\begin{Vmatrix}a_1 & a_2 & a_3\\ b_1 & b_2 & b_3\\ c_1 & c_2 & c_3\end{Vmatrix}, \begin{Vmatrix}\tilde a_1 & \tilde a_2 & \tilde a_3\\ \tilde b_1 & \tilde b_2 & \tilde b_3\\ \tilde c_1 & \tilde c_2 & \tilde c_3\end{Vmatrix}\right)$:

$$\tilde g = \alpha^2[(\omega_1+\tilde\omega_2)^2 + (\tilde\omega_1+\omega_2)^2] + \beta^2[(\omega_1-\tilde\omega_2)^2 + (\tilde\omega_1-\omega_2)^2] + \gamma^2(\omega_3+\tilde\omega_3)^2,$$

where
$$\omega_1 = a_2 da_3 + b_2 db_3 \pm c_2 dc_3$$
$$\omega_2 = a_3 da_1 + b_3 db_1 \pm c_3 dc_1$$
$$\omega_3 = a_1 da_2 + b_1 db_2 \pm c_1 dc_2$$
and $\tilde\omega_1$, $\tilde\omega_2$, $\tilde\omega_3$ are given by similar expressions in $\tilde a_i$, $\tilde b_i$, $\tilde c_i$, $d\tilde a_i$, $d\tilde b_i$, $d\tilde c_i$.

Here $\alpha, \beta, \mu$ are positive real parameters and the (+) and (-) signs in $\omega_1$, $\omega_2$, $\omega_3$ correspond to the elliptic case 5a and to the hyperbolic case 5b respectively.

The typical symmetry at the origin of $M$ is induced by the following transformation of $GL(3,R)\times GL(3,R)$ :

$$\left(\begin{Vmatrix}a_1 & a_2 & a_3\\ b_1 & b_2 & b_3\\ c_1 & c_2 & c_3\end{Vmatrix}, \begin{Vmatrix}\tilde a_1 & \tilde a_2 & \tilde a_3\\ \tilde b_1 & \tilde b_2 & \tilde b_3\\ \tilde c_1 & \tilde c_2 & \tilde c_3\end{Vmatrix}\right) \longmapsto \left(\begin{Vmatrix}\tilde a_1 & -\tilde a_2 & -\tilde a_3\\ -\tilde b_1 & \tilde b_2 & \tilde b_3\\ -\tilde c_1 & \tilde c_2 & \tilde c_3\end{Vmatrix}, \begin{Vmatrix}a_1 & -a_2 & a_3\\ -b_1 & b_2 & -b_3\\ c_1 & -c_2 & c_3\end{Vmatrix}\right)$$

Types 6a, 6b.

The underlying homogeneous space is either $SU(3)/SU(2)$, or $SU(2,1)/SU(2)$.

$M$ is the submanifold of $C^3(z^1,z^2,z^3)$ given by the relation $z^1\bar{z}^1 + z^2\bar{z}^2 \pm z^3\bar{z}^3 = \pm 1$. The Riemann metric on $M$ is induced by the following Hermitian metric on $C^3$:

$$\tilde{g} = \lambda(dz^1 d\bar{z}^1 + dz^2 d\bar{z}^2 \pm dz^3 d\bar{z}^3) +$$

$$+ \mu(z^1 d\bar{z}^1 + z^2 d\bar{z}^2 \pm z^3 d\bar{z}^3)(\bar{z}^1 dz^1 + \bar{z}^2 dz^2 \pm \bar{z}^3 dz^3)$$

where $\lambda$, $\mu$ are real parameters such that $\lambda > 0$, $\mu \neq 0$ and $\mu \pm \lambda > 0$. The (+) and (-) signs correspond to the elliptic case 6a and to the hyperbolic case 6b respectively.

The typical symmetry at the point $(0,0,1)$ of $M$ is induced by the following transformation of $C^3$: $z^{1'} = \bar{z}^2$, $z^{2'} = -\bar{z}^1$, $z^{3'} = \bar{z}^3$.

Type 7.

The underlying homogeneous space is the real matrix group:

$$\begin{Vmatrix} e^{\lambda t} & 0 & 0 & 0 & x \\ 0 & e^{-\lambda t} & 0 & 0 & y \\ te^{\lambda t} & 0 & e^{\lambda t} & 0 & u \\ 0 & -te^{-\lambda t} & 0 & e^{-\lambda t} & v \\ 0 & 0 & 0 & 0 & 1 \end{Vmatrix}$$

($t,x,y,u,v$ are real variables and $\lambda$ is a real parameter).

$(M,g)$ is the space $R^5(x,y,u,v,t)$ with a Riemann metric

$$g = (dt)^2 + e^{-2\lambda t}(tdx - du)^2 + e^{2\lambda t}(tdy + dv)^2 + \mu(e^{-2\lambda t}dx^2 + e^{2\lambda t}dy^2) +$$

$$+ 2\gamma(dydu - dxdv),$$

where $\lambda$, $\mu$, $\gamma$ are real parameters, $\lambda \geq 0$, $\mu > 0$, $\gamma^2 < \mu$.

The typical symmetry at the point $(0,\dots,0)$ is the transformation $x' = -y$, $y' = x$, $u' = -v$, $v' = u$, $t' = -t$.

Types 8a, 8b.

The underlying homogeneous space is either $I^e(R^3)/SO(2)$, or $I^h(R^3)/$ $/SO(2)$, where $I^e(R^3)$, or $I^h(R^3)$, denotes the group of all positive affine transformations of the space $R^3(x,y,z)$ preserving the differential form $dx^2 + dy^2 + dz^2$, or $dx^2 + dy^2 - dz^2$, respectively. ($I^e(R^3)$ is the semidirect product of $SO(3)$ and $t(3)$, and $I^h(R^3)$ is the semidirect product of $SO(2,1)$ and $t(3)$, where $t(3)$ denotes the translation group of $R^3$.)

$M$ is the submanifold of $R^6(x,y,z;\alpha,\beta,\gamma)$ given by the relation $\alpha^2+\beta^2\pm\gamma^2=\pm1$. The Riemann metric on $M$ is induced by the following regular invariant quadratic form on $R^6$:

$$g = dx^2 + dy^2 \pm dz^2 + \lambda^2(d\alpha^2 + d\beta^2 \pm d\gamma^2) + [\mu^2 \pm (-1)](\alpha dx + \beta dy \pm \gamma dz)^2$$

where $\lambda > 0$, $\mu > 0$ are real parameters. The (+) and (-) signs correspond to the elliptic case 8a and to the hyperbolic case 8b respectively.

The typical symmetry at the point $(0,0,0;0,0,1)$ of $M$ is induced by the following transformation of $R^6$:
$x' = -y,\quad y' = x,\quad z' = -z,\quad \alpha' = \beta,\quad \beta' = -\alpha,\quad \gamma' = \gamma.$

Type 9. (Spaces of order 6.)

The underlying homogeneous space is the matrix group:

$$\left\|\begin{matrix} e^{-(u+v)} & 0 & 0 & x \\ 0 & e^u & 0 & y \\ 0 & 0 & e^v & z \\ 0 & 0 & 0 & 1 \end{matrix}\right\| .$$

$(M,g)$ is the space $R^5(x,y,z,u,v)$ with a Riemann metric

$$g = a^2(du^2 + dv^2 + dudv) + (b^2+1)(e^{2(u+v)}dx^2 + e^{-2u}dy^2 + e^{-2v}dz^2) + (b^2-2)(e^v dxdy + e^u dxdz - e^{-(u+v)}dydz), \quad \text{where } a > 0,\ b > 0.$$

The typical symmetry at the point $(0,\ldots,0)$ is the transformation $x' = y,\quad y' = -z,\quad z' = x;\quad u' = v,\quad v' = -(u+v).$

-------

Outline of the proof. The next assertion is left to the reader:

Proposition VI.6. The essentially different elements of $\mathfrak{D}^{*5}$ are the following:

a) $(i, -i, i, -i, -1)$
b) $(\theta, \theta^2, \theta^3, \theta^4, \theta^5)$, $\quad \theta = e^{2\pi i/6}$
c) $(e^{\pi i/4}, e^{-\pi i/4}, i, -i, -1)$
d) $(i, -i, -1, -1, -1)$
e) $(-1, -1, -1, -1, -1)$.

Now, the system of eigenvalues e) yields only Riemannian symmetric spaces. As for the case d), we obtain only reducible infinitesimal s-manifolds and thus reducible generalized symmetric spaces. (The details are left to the reader.)

Let us consider the case c). Let $V$ be a 5-dimensional vector space over $R$, and let $S: V^c \longrightarrow V^c$ be a real transformation with the eigenvalues $e^{\pi i/4}$, $e^{-\pi i/4}$, $i$, $-i$, $-1$. Let $U_1$, $\bar{U}_1$, $U_2$, $\bar{U}_2$, $W$ be the corresponding eigenvectors (where $W \in V$ is real). For each skew-symmetric tensor $\tilde{T} \neq 0$ of type (1,2) such that $\tilde{T}(SZ,SZ') = S(\tilde{T}(Z,Z'))$ we can put

$$\tilde{T}(U_2,W) = \alpha\bar{U}_2, \quad \tilde{T}(U_1,\bar{U}_2) = \beta\bar{U}_1, \quad \tilde{T}(\bar{U}_2,W) = \bar{\alpha}U_2, \quad \tilde{T}(\bar{U}_1,U_2) = \bar{\beta}U_1,$$

where $\alpha$, $\beta$ are complex numbers (not both equal to zero) and $\tilde{T}(Z,Z') = 0$ for any other combination of eigenvectors $Z,Z'$. Suppose that $(V,g,S,\tilde{R},\tilde{T})$ be an infinitesimal s-manifold with $S$ and $\tilde{T}$ as above. Notice that $g(U_1,\bar{U}_1) > 0$, $g(U_2,\bar{U}_2) > 0$, $g(W,W) > 0$. Supposing for a moment that $\alpha\beta \neq 0$, we show easily that the Lie algebra $\underline{k}$ is a zero algebra; thus $\tilde{R} = 0$. The first Bianchi identity $\mathfrak{S}[\tilde{T}(\tilde{T}(U_2,W),U_1)] = 0$ implies $\alpha\beta\bar{U}_1 = 0$, and hence $\alpha = 0$ or $\beta = 0$. In the first case we can replace $S$ by a transformation $S'$ with the eigenvalues sub a). In the second case we can replace $S$ by a transformation $S''$ with the eigenvalues sub b). Thus we can limit our investigations only to the cases a) and b).

Now, the classification in the case a) is based on the following

<u>Proposition VI.7</u>. Let $(V,g,S,\tilde{R},\tilde{T})$ be a 5-dimensional infinitesimal s-manifold with a system of eigenvalues $(i,-i,i,-i,-1)$. Denote by $V^{(i)}$, $V^{(-i)}$, $V^{(-1)}$ the corresponding eigenspaces of $S$ in $V^c$, so that $V^c = V^{(i)} + V^{(-i)} + V^{(-1)}$.
Then for any basis $\{U_1,U_2\}$ of $V^{(i)}$ and any $W \in V^{(-1)} \cap V$ the tensor $\tilde{T}$ satisfies the relations

$$\tilde{T}(U_1,U_2) = \mu W, \qquad \tilde{T}(U_i,\bar{U}_j) = 0, \qquad i,\ j = 1,2.$$

Further, the partial map $\tilde{T}: V^{(i)} \times V^{(-1)} \longrightarrow V^{(-i)}$ reduces, for a suitable choice of $U_1, U_2 \in V^{(i)}$, $W \in V^{(-1)} \cap V$, to one and only one of the following canonical forms:

1) $\tilde{T}(U_i,W) = 0$, $\quad i = 1,2$
2) $\tilde{T}(U_1,W) = \lambda_1\bar{U}_1$, $\tilde{T}(U_2,W) = \lambda_2\bar{U}_2$, $\quad \lambda_1 > 0$, $\lambda_2 \geq 0$, $\lambda_1 \geq \lambda_2$
3) $\tilde{T}(U_1,W) = \lambda\bar{U}_2$, $\tilde{T}(U_2,W) = \bar{\lambda}\bar{U}_1$, $\quad \lambda$ imaginary
4) $\tilde{T}(U_1,W) = \lambda\bar{U}_1 + \bar{U}_2$, $\tilde{T}(U_2,W) = \lambda\bar{U}_2$, $\quad \lambda \geq 0$ real.

Now: In the case 1) we obtain Type 1 of our classification list.
In the case 2) we obtain Types 2 and 3.
In the case 3) we obtain Types 4, 5a, 5b, 6a and 6b.

Finally, in the case 4) we obtain Types 7, 8a and 8b. The calculations are rather long and not always routine. The reader can see the booklet [K3] for details.

The classification in the case b) (a system of eigenvalues of order 6) gives the remaining Type 9.

Remark VI.8. It can be shown, by very long calculations, that the parameters in each family are "infinitesimal" invariants. On the other hand, it is not difficult to show that the spaces of different types are always mutually non-isometric.

## D i m e n s i o n $n = 6$ .

Recently, E.Kurcius (Ph.D.Thesis, University of Katowice 1978) has proved that the proper, simply connected and irreducible generalized symmetric Riemannian spaces of dimension $n = 6$ can be classified in the gross as follows:

A) Spaces of order 6 (which are closely related to Type 9 of the classification in the dimension $n = 5$).

B) Spaces of order 4.

C) Spaces of order 3.

However, the problem is still far from its final solution.

---

References: [K1],[K3], and [TL1] (with incomplete results).

CHAPTER VII

# THE CLASSIFICATION OF GENERALIZED AFFINE SYMMETRIC SPACES IN LOW DIMENSIONS

M.Berger, [B], has worked out a complete list of local structures of affine symmetric spaces admitting a transitive semi-simple group of automorphisms. He sets aside the spaces of "solvable" and "mixed" type; for such spaces only a topological structural theorem has been proved.

As Proposition V.9 suggests, in the case of generalized affine symmetric spaces the solvable groups play even more important part then in the classical situation. (See also Note 1.) Hence we can guess that the classification of local structures of g.a.s. spaces is a very difficult problem.

In this paragraph we shall describe the classification procedure for the low dimensions. The method is similar to that used in Chapter VI, and we shall limit ourselves only to the explanation of some of the differences.

A) As in the Riemannian case, we start from the classification of "distinguished" systems of eigenvalues. According to Theorem V.16, we have to describe the set $\mathfrak{E}^n$, i.e., to find all minimal $\theta$-varieties which are contained in $\mathcal{A}^n$.

In the Riemannian, or unitary affine case, all minimal $\theta$-varieties are finite sets. In the general case, some minimal $\theta$-varieties may be also 1-parametric families - they are those describing g.a.s. spaces of infinite order. (It is not known to the author if there are also higher-dimensional minimal $\theta$-varieties, cf. Problem V.35.) Yet, the number of distinct minimal $\theta$-varieties is still finite.

B) Starting from an element $(\theta_1,\dots,\theta_n)$ of $\mathfrak{E}^n$ we are seeking canonical types of all infinitesimal s-manifolds $(V,S,\tilde{R},\tilde{T})$ such that $S$ is semi-simple and has the eigenvalues $\theta_i$. As a rule, we get much more different canonical types than in the Riemannian case. (It corresponds to the fact that the full linear group acting on a tensor space admits usually more types of orbits than the orthogonal group acting on the same space.)

C) The notion "irreducible" in the Riemannian case is replaced here by the notion "primitive". According to Proposition IV.13, a g.a.s. space is primitive if and only if it does not come from a reducible regular s-manifold. A simply connected g.a.s. space is primitive if and only if it does not come from a reducible infinitesimal s-manifold. (Proposition IV.16.) Finally, according to Theorem IV.43 we see that infinitesimal s-manifolds with reducible systems of eigenvalues are always reducible. Because we are interested only in primitive g.a.s. spaces, we can limit ourselves, in the point A), to the subset $\mathcal{E}^{*n} \subset \mathcal{E}^n$ consisting of irreducible systems $(\theta_1,\dots,\theta_n)$ of eigenvalues.

-------

In the following we shall present the classification list by S.Wegrzynowski of all local structures of primitive g.a.s. spaces of dimensions 3, 4 which are not locally symmetric. Each g.a.s. space in question (or, more strictly, each class of locally isomorphic g.a.s. spaces) is represented here by a prime regular homogeneous s-manifold $(G,H,\mathfrak{S})$, where the system of eigenvalues of $\mathfrak{S}$ is of the form $(\theta_1,\dots,\theta_n,\underbrace{1,\dots,1}_{\dim H})$, $(\theta_1,\dots,\theta_n) \in \mathcal{E}^n$. (See [W1].)

(There is one exception concerning the type III in dimension 4, where the model is given only as a local regular s-triplet $(\underline{g},\underline{h},\nu)$. The reason for this is that the corresponding Lie groups do not look too nice if we use the most routine representation by the affine transformations of the space $R^4$. But it may be interesting to look for some more natural geometric representations in this situation.)

## The Classification List.

Dimension n = 3.

There are exactly three types of generalized affine symmetric spaces denoted as I, II, III. Here I is of order 4, and II, III are of infinite order. All these spaces are primitive and they are represented by the following matrix groups and their automorphisms:

(I):

$$G_I = \begin{Vmatrix} e^{-z} & 0 & x \\ 0 & e^z & y \\ 0 & 0 & 1 \end{Vmatrix}$$

$\mathfrak{S}$: $x' = -y, \quad y' = x, \quad z' = -z.$

(II):

$$G_{II} = \begin{Vmatrix} \cos z & -\sin z & x \\ \sin z & \cos z & y \\ 0 & 0 & 1 \end{Vmatrix}$$

$\sigma$: $x' = -\alpha x$, $y' = \alpha y$, $z' = -z$, $(\alpha \neq 0, \pm 1)$.

(III):

$$G_{III} = \begin{Vmatrix} 1 & x & y \\ 0 & 1 & z \\ 0 & 0 & 1 \end{Vmatrix}$$

$\sigma$ : $x' = -\alpha x$, $y' = \alpha y$, $z' = -z$, $(\alpha \neq 0, \pm 1)$.

Dimension $n = 4$.

The complete classification is given by three types (I, II and III below) of generalized affine symmetric spaces of order 3, two types (IV and V) of order 4 and 11 types (VI - XVI) of infinite order.

All these spaces are primitive and they are represented by the following regular homogeneous s-manifolds $(G, H, \sigma)$:

A. Generalized affine symmetric spaces of order 3.

(I):

$$G_1 = \begin{Vmatrix} a & b & e \\ c & d & f \\ 0 & 0 & 1 \end{Vmatrix}, \quad ad - bc = 1,$$

$$H_1 = \begin{Vmatrix} \cos t & -\sin t & 0 \\ \sin t & \cos t & 0 \\ 0 & 0 & 1 \end{Vmatrix},$$

$\sigma$: $a' = \frac{1}{4}a + \frac{\sqrt{3}}{4}b + \frac{\sqrt{3}}{4}c + \frac{3}{4}d$, $b' = -\frac{\sqrt{3}}{4}a + \frac{1}{4}b - \frac{3}{4}c + \frac{\sqrt{3}}{4}d$,

$c' = -\frac{\sqrt{3}}{4}a + \frac{3}{4}b + \frac{1}{4}c + \frac{\sqrt{3}}{4}d$, $d' = \frac{3}{4}a - \frac{\sqrt{3}}{4}b - \frac{\sqrt{3}}{4}c + \frac{1}{4}d$,

$e' = -\frac{1}{2}e - \frac{\sqrt{3}}{2}f$, $f' = \frac{\sqrt{3}}{2}e - \frac{1}{2}f$.

(II):

$$G_2 = \begin{Vmatrix} e^s\cdot\cosh t & e^s\cdot\sinh t & 0 & a \\ e^s\cdot\sinh t & e^s\cdot\cosh t & 0 & b \\ 0 & 0 & e^{-2s} & c \\ 0 & 0 & 0 & 1 \end{Vmatrix}$$

(cosh, sinh denote hyperbolic functions.)

$$H_2 = \begin{Vmatrix} 1 & 0 & 0 & -a \\ 0 & 1 & 0 & 0 \\ 0 & 0 & 1 & 2a \\ 0 & 0 & 0 & 1 \end{Vmatrix},$$

$\sigma$: $a' = \frac{1}{2}a - \frac{1}{2}b - \frac{1}{4}c$, $b' = \frac{1}{2}a - \frac{1}{2}b + \frac{1}{4}c$, $c' = -2a - 2b$, $u' = v$, $v' = -u + v$, where $u = s + t$, $v = s - t$.

(III): The local regular s-triplet $(\underline{g}_3, \underline{h}_3, \nu)$ is defined as follows:

$\underline{g}_3$ is isomorphic to a 7-dimensional Lie algebra of infinitesimal transformations of the space $R^4[x,y,u,v]$ with the basis

$$X_1 = \frac{\partial}{\partial x}, \quad X_2 = \frac{\partial}{\partial y}, \quad X_3 = (-2x + u)\frac{\partial}{\partial x} + (2y - v)\frac{\partial}{\partial y} + 3x\frac{\partial}{\partial u} - 3y\frac{\partial}{\partial v},$$

$$X_4 = (2y + v)\frac{\partial}{\partial x} + (2x + u)\frac{\partial}{\partial y} + 3y\frac{\partial}{\partial u} + 3x\frac{\partial}{\partial v},$$

$$A = -y\frac{\partial}{\partial x} + x\frac{\partial}{\partial y} + 3v\frac{\partial}{\partial u} - 3u\frac{\partial}{\partial v}, \quad B = \frac{\partial}{\partial u}, \quad C = \frac{\partial}{\partial v}.$$

$\underline{h}_3$ corresponds to the 3-dimensional subalgebra $(A, B, C)$, and $\nu$ is an automorphism of order 3 of $\underline{g}_3$ induced by the following transformation of $R^4$:

$$x' = \frac{\sqrt{3}}{2}x + \frac{1}{2}y, \quad y' = -\frac{1}{2}x + \frac{\sqrt{3}}{2}y, \quad u' = u, \quad v' = v.$$

B. Generalized affine symmetric spaces of order 4.

(IV):

$$G_4 = \begin{Vmatrix} \cosh t & -\sinh t & 0 & 0 & a \\ -\sinh t & \cosh t & 0 & 0 & b \\ 0 & 0 & \cosh \varkappa t & \frac{1}{\varkappa}\sinh \varkappa t & c \\ 0 & 0 & \varkappa\sinh \varkappa t & \cosh \varkappa t & d \\ 0 & 0 & 0 & 0 & 1 \end{Vmatrix} \quad (\varkappa > 0).$$

$$H_4 = \begin{Vmatrix} 1 & 0 & 0 & 0 & 0 \\ 0 & 1 & 0 & 0 & 0 \\ 0 & 0 & 1 & 0 & 0 \\ 0 & 0 & 0 & 1 & d \\ 0 & 0 & 0 & 0 & 1 \end{Vmatrix}$$

$\sigma$: $t' = -t$, $a' = -b$, $b' = a$, $c' = -c$, $d' = d$.

(V):

$$G_5 = \begin{Vmatrix} \cosh t & -\sinh t & 0 & 0 & a \\ -\sinh t & \cosh t & 0 & 0 & b \\ 0 & 0 & \cos \varkappa t & \frac{1}{\varkappa}\sin \varkappa t & c \\ 0 & 0 & -\varkappa\sin \varkappa t & \cos \varkappa t & d \\ 0 & 0 & 0 & 0 & 1 \end{Vmatrix} \quad (\varkappa > 0).$$

$H_5 = H_4$,

$\sigma$: $t' = -t$, $a' = -b$, $b' = a$, $c' = -c$, $d' = d$.

C. Generalized affine symmetric spaces of infinite order.

(VI):

$$G_6 = \begin{Vmatrix} 1 & t & \frac{1}{2}t^2 & a \\ 0 & 1 & t & b \\ 0 & 0 & 1 & c \\ 0 & 0 & 0 & 1 \end{Vmatrix}, \quad H_6 = \mathrm{Id}.$$

$\sigma$: $a' = \alpha^3 a$, $b' = \alpha b$, $c' = \frac{1}{\alpha}c$, $t' = \alpha^2 t$ $(\alpha \neq 0, \pm 1)$.

(VII):

$$G_7 = \begin{Vmatrix} \cosh t & \sinh t & 0 & a \\ \sinh t & \cosh t & 0 & b \\ b\cdot\cosh t - a\cdot\sinh t & b\cdot\sinh t - a\cdot\cosh t & 1 & c \\ 0 & 0 & 0 & 1 \end{Vmatrix}$$

$H_7 = \mathrm{Id}$,

$\sigma$: $t' = -t$, $a' = \alpha a$, $b' = -\alpha b$, $c' = -\alpha^2 c$, $(\alpha \neq 0, \pm 1)$.

(VIII):

$$G_8 = \begin{Vmatrix} \cos t & \sin t & 0 & a \\ -\sin t & \cos t & 0 & b \\ b\cdot\cos t + a\cdot\sin t & b\cdot\sin t - a\cdot\cos t & 1 & c \\ 0 & 0 & 0 & 1 \end{Vmatrix}$$

$H_8 = \mathrm{Id}$,

$\sigma$: $t' = -t$, $a' = \alpha a$, $b' = -\alpha b$, $c' = -\alpha^2 c$, $(\alpha \neq 0, \pm 1)$.

(IX):

$$G_9 = \begin{Vmatrix} a & b & e \\ c & d & f \\ 0 & 0 & 1 \end{Vmatrix}, \quad ad - bc = 1,$$

$$H_9 = \begin{Vmatrix} e^{-t} & 0 & 0 \\ 0 & e^{t} & 0 \\ 0 & 0 & 1 \end{Vmatrix},$$

$\sigma$: $a' = a$, $b' = \frac{1}{\alpha^2}b$, $c' = \alpha^2 c$, $d' = d$, $e' = \frac{1}{\alpha}e$, $f' = \alpha f$,
$(\alpha \neq 0, \pm 1)$.

(X):

$$G_{10} = \begin{Vmatrix} 1 & a & b & c \\ 0 & 1 & d & e \\ 0 & 0 & 1 & d \\ 0 & 0 & 0 & 1 \end{Vmatrix} .$$

$$H_{10} = \begin{Vmatrix} 1 & 0 & b & 0 \\ 0 & 1 & 0 & 0 \\ 0 & 0 & 1 & 0 \\ 0 & 0 & 0 & 1 \end{Vmatrix} .$$

$\sigma$ : $a' = \frac{1}{\alpha}a$, $b' = b$, $c' = \alpha c$, $d' = \alpha d$, $e' = \alpha^2 e$, $(\alpha \neq 0, \pm 1)$.

(XI):

$$G_{11} = \begin{Vmatrix} 1 & t & 0 & 0 & a \\ 0 & 1 & 0 & 0 & b \\ 0 & 0 & \cos t & \sin t & c \\ 0 & 0 & \sin t & \cos t & d \\ 0 & 0 & 0 & 0 & 1 \end{Vmatrix}$$

$$H_{11} = \begin{Vmatrix} 1 & 0 & 0 & 0 & 0 \\ 0 & 1 & 0 & 0 & 0 \\ 0 & 0 & 1 & 0 & 0 \\ 0 & 0 & 0 & 1 & d \\ 0 & 0 & 0 & 0 & 1 \end{Vmatrix} = H_4$$

$\sigma$ : $t' = -t$, $a' = \alpha a$, $b' = -\alpha b$, $c' = -c$, $d' = d$, $(\alpha \neq 0, \pm 1)$.

(XII):

$$G_{12} = \begin{Vmatrix} 1 & t & 0 & 0 & a \\ 0 & 1 & 0 & 0 & b \\ 0 & 0 & \cosh t & \sinh t & c \\ 0 & 0 & \sinh t & \cosh t & d \\ 0 & 0 & 0 & 0 & 1 \end{Vmatrix}$$

$H_{12} = H_{11}$,

$\sigma$ : $t' = -t$, $a' = \alpha a$, $b' = -\alpha b$, $c' = -c$, $d' = d$, $(\alpha \neq 0, \pm 1)$.

(XIII):

$$G_{13} = \begin{Vmatrix} \cos t & \sin t & 0 & 0 & a \\ -\sin t & \cos t & 0 & 0 & b \\ 0 & 0 & \cos \varkappa t & \frac{1}{\varkappa}\sin \varkappa t & c \\ 0 & 0 & -\varkappa \sin \varkappa t & \cos \varkappa t & d \\ 0 & 0 & 0 & 0 & 1 \end{Vmatrix} \qquad (\varkappa > 0).$$

$H_{13} = H_{11}$,

$\sigma$ : $t' = -t$, $a' = \alpha a$, $b' = -\alpha b$, $c' = -c$, $d' = d$, $(\alpha \neq 0, \pm 1)$.

(XIV):

$$G_{14} = \begin{Vmatrix} \cos t & \sin t & 0 & 0 & a \\ -\sin t & \cos t & 0 & 0 & b \\ 0 & 0 & \cosh \varkappa t & \frac{1}{\varkappa}\sinh \varkappa t & c \\ 0 & 0 & \varkappa \sinh \varkappa t & \cos \varkappa t & d \\ 0 & 0 & 0 & 0 & 1 \end{Vmatrix} \quad (\varkappa > 0).$$

$H_{14} = H_{11}$,

$\sigma$: $t' = -t$, $a' = \alpha a$, $b' = -\alpha b$, $c' = -c$, $d' = d$, $(\alpha \neq 0, \pm 1)$.

(XV):

$$G_{15} = \begin{Vmatrix} 1 & t & -\frac{1}{2}t^2 & \frac{1}{6}t^3 & a \\ 0 & 1 & -t & \frac{1}{2}t^2 & b \\ 0 & 0 & 1 & -t & c \\ 0 & 0 & 0 & 1 & d \\ 0 & 0 & 0 & 0 & 1 \end{Vmatrix}$$

$$H_{15} = \begin{Vmatrix} 1 & 0 & 0 & 0 & 0 \\ 0 & 1 & 0 & 0 & b \\ 0 & 0 & 1 & 0 & 0 \\ 0 & 0 & 0 & 1 & 0 \\ 0 & 0 & 0 & 0 & 1 \end{Vmatrix}$$

$\sigma$: $t' = \frac{1}{\alpha}t$, $a' = \frac{1}{\alpha}a$, $b' = b$, $c' = \alpha c$, $d' = \alpha^2 d$, $(\alpha \neq 0, \pm 1)$.

(XVI):

$$G_{16} = \begin{Vmatrix} 1 & -t & -\frac{1}{2}t^2 & \frac{1}{6}t^3 & a \\ 0 & 1 & t & -\frac{1}{2}t^2 & b \\ 0 & 0 & 1 & -t & c \\ 0 & 0 & 0 & 1 & d \\ 0 & 0 & 0 & 0 & 1 \end{Vmatrix}$$

$$H_{16} = \begin{Vmatrix} 1 & 0 & 0 & 0 & 0 \\ 0 & 1 & 0 & 0 & 0 \\ 0 & 0 & 1 & 0 & c \\ 0 & 0 & 0 & 1 & 0 \\ 0 & 0 & 0 & 0 & 1 \end{Vmatrix},$$

$\sigma$: $t' = \alpha t$, $a' = \alpha^2 a$, $b' = \alpha b$, $c' = c$, $d' = \frac{1}{\alpha}d$, $(\alpha \neq 0, \pm 1)$.

<u>Outline of the proof</u>.

A) <u>dim M = 3</u>.

First we prove that all minimal $\theta$-varieties composing the set $\mathfrak{C}^{*3}$ are represented by the following elements:

(1) $(i, -i, -1)$
(2) $(\alpha, -\alpha, -1)$
(3) $(\alpha^2, \alpha, \frac{1}{\alpha})$
(4) $(\alpha^2, \alpha, \alpha)$
$\Big\}\ \alpha \neq 0, 1, -1$ (for (2)–(4))
(5) $(-1, -1, -1)$

Then by the procedure as above we derive Type I from the element (1), Type II from the element (2) and Type III from each of the elements (2), (3) and (4). The element (5) yields only affine symmetric spaces.

B) <u>dim M = 4</u>.

The minimal $\theta$-varieties composing the set $\mathcal{C}^{*4}$ are either finite sets or 1-parametric families. The finite sets are represented by the elements

a) $(\theta, \theta^2, \theta, \theta^2)$, $\quad \theta = e^{2\pi i/3}$
b) $(\theta, \theta^2, \theta^3, \theta^4)$, $\quad \theta = e^{2\pi i/5}$
c) $(i, -i, -1, -1)$
d) $(-1, -1, -1, -1)$

exactly as in the Riemannian case. Here the element a) gives the types I, II, III, element c) the types IV, V, and also I, b) only the type I, and d) yields only affine symmetric spaces.

The 1-parametric families are represented by the elements

(1) $(\alpha^2, \alpha, \frac{1}{\alpha}, \frac{1}{\alpha^2})$, $\quad \alpha \neq 0, 1, -1$

(2) $(\alpha, \alpha, \frac{1}{\alpha^2}, \frac{1}{\alpha})$

(3) $(\alpha, \alpha, \alpha^2, \frac{1}{\alpha})$

(4) $(\alpha, -\alpha, -1, -1)$

(5) $(\alpha, \alpha^2, \alpha^3, \frac{1}{\alpha})$

(6) $(-1, \alpha, -\alpha, -\alpha^2)$

(7) $(-1, \alpha, -\alpha, \frac{1}{\alpha})$

(8) $(\alpha, \alpha^3, \frac{1}{\alpha}, \frac{1}{\alpha^2})$

(9) $(-1, \alpha, -\alpha, -\alpha)$

(10) $(\alpha, \alpha, \alpha^2, \alpha^3)$

(11) $(-1, -1, \alpha, -\frac{1}{\alpha})$

(12) $(\alpha, \frac{1}{\alpha}, \alpha^2, \alpha^2)$

(13) $(\alpha, \alpha, \alpha, \alpha^2)$

(14) $(\alpha, \alpha, \alpha^2, \alpha^2)$.

Here (1) yields type IX, (2) yields type XV, (3) yields types X, XVI and also IX, (4) yields types XI, XII, XIII, XIV, and also types IX, and IV, V. (5) yields type VI, and also IX, (6) yields types VI, VII, VIII, (7) yields types VI and IX. (8), (9) and (10) yield only type VI. (11), (12), (13) and (14) yield reducible s-manifolds only.

The details are given in [W2].

---

References: [B],[K8],[W1],[W2].

## Note 1. Existence of generalized symmetric spaces of solvable type.

In this Note we shall present two existence theorems (for the Riemannian and the affine case) proved by M.Božek ([Bo1],[Bo2]).

**Theorem 1.** For every even integer $m \geq 4$ there is an irreducible generalized symmetric Riemannian space of order $m$ which is diffeomorphic to $R^{m-1}$ and such that the identity component of its full isometry group is solvable.

**Theorem 2.** For every integer $m \geq 3$ there is a primitive generalized affine symmetric space of order $m$ which is diffeomorphic to $R^{2m-2}$ and such that the identity component of the full affine group is solvable.

### Construction and the outline of the proof.

A) The Riemannian case.

Let $n \geq 1$ be an integer, denote by $G_n$ the matrix group consisting of all matrices of the form

$$\begin{Vmatrix} e^{u_o} & 0 & \ldots & 0 & x_0 \\ 0 & e^{u_1} & \ldots & 0 & x_1 \\ \ldots & \ldots & \ldots & \ldots & \ldots \\ 0 & 0 & \ldots & e^{u_n} & x_n \\ 0 & 0 & \ldots & 0 & 1 \end{Vmatrix}$$

where $(x_0,x_1,\ldots,x_n,u_1,\ldots,u_n) \in R^{2n+1}$ is arbitrary, and $u_o = -u_1-\ldots-u_n$. Thus the underlying manifold for $G_n$ is $R^{2n+1}$ and the group $G_n$ is solvable. Let us consider a left-invariant metric $g$ on $G_n$ defined by

$$g = \sum_{i=o}^{n} e^{-2u_i}(dx_i)^2 + a \sum_{\alpha,\beta=1}^{n} du_\alpha du_\beta \; ; \qquad a > 0.$$

For $n = 1$, the corresponding Riemannian manifolds are nothing but the 3-dimensional generalized symmetric spaces from Theorem VI.2. For $n = 2$ we obtain a 1-parametric family of the 5-dimensional generalized symmetric spaces of Type 9, Theorem VI.5.

Let us consider the automorphism of $G_n$ given by the map $s: (x_0,x_1,\dots,x_n,u_1,\dots,u_n) \mapsto (-x_n,x_0,x_1,\dots,x_{n-1},u_0,u_1,\dots,u_{n-1})$, where $u_0$ again denotes the number $-(u_1+\dots+u_n)$. The map $s$ is also an isometry of $(G_n,g)$ with the identity element as a unique fixed point, and we have $s^{2n+2} = \mathrm{id}$. Hence the Riemannian manifold $(G_n,g)$ is a $(2n+2)$-symmetric space.

Denote by $\tilde{\nabla}$ the Cartan $(-)$ connection on the group $G_n$ and by $\nabla$ the Riemannian connection of $g$. Then the difference tensor field $D = \nabla - \tilde{\nabla}$ is invariant with respect to the left translations of $G_n$. Now, for the covariant derivative of the curvature tensor $R$ we have $\nabla^k R = D^k R$ for $k = 0,1,\dots$ .

Let $I(G_n,e)$ denote the group of all isometries of $(G_n,g)$ leaving the identity element $e$ fixed. Because our Riemannian space is analytic, complete and simply connected, the elements of $I(G_n,e)$ are in a (1-1)-correspondence with the linear transformations $S$ of the tangent space $V = (G_n)_e$ such that $S(g) = g$, $S(R) = R$, and $S(D^kR) = D^kR$ for $k = 1,2,\dots$ . Hence the elements of the Lie algebra of $I(G_n,g)$ are in a (1-1)-correspondence with the endomorphisms $A$ of $V$ which, as derivations of the tensor algebra $\mathcal{T}(V)$, satisfy $A(g) = A(R) = A(DR) = \dots = A(D^kR) = \dots = 0$. Now, by a rather lenghty calculation we derive that $A(g) = A(R) = A(DR) = 0$ implies $A = 0$. Hence the group $I(G_n,e)$ is finite and the identity component of the full isometry group $I(G_n)$ is isomorphic to $G_n$.

For $\varphi \in I(G_n,e)$, the map $g \mapsto \varphi \circ g \circ \varphi^{-1}$ is an automorphism of $G_n$ onto itself and thus the tangent map $\varphi_{*e}$ is an isometric automorphism of the Lie algebra $\underline{g}_n$, where we identity $\underline{g}_n$ with $T_e(G_n)$. We can see that the map $\varphi \mapsto \varphi_{*e}$ is an isomorphism of $I(G_n,e)$ onto the group $L$ of all isometric automorphisms of $\underline{g}_n = T_e(G_n)$. It remains to find the structure of the group $L$.

Let us denote
$$X_i = e^{u_i}\frac{\partial}{\partial x_i}, \qquad i = 0,1,\dots,n$$
$$U_\alpha = \frac{\partial}{\partial u_\alpha}, \qquad \alpha = 1,\dots,n .$$

A direct calculation shows that $X_0,X_1,\dots,X_n,U_1,\dots,U_n$ form a basis of the Lie algebra $\underline{g}_n$ of the group $G_n$. Now, let $Z_2$ denote the multiplicative group with two elements $1, -1$, and $S_{n+1}$ be the symmetric group of the index set $\{0,1,\dots,n\}$. Consider the map

$\Phi : (Z_2)^{n+1} \times S_{n+1} \longrightarrow GL(\underline{g}_n)$ given by the formulas

$\Phi(\varepsilon, \sigma) X_i = \varepsilon_i X_{\sigma(i)}$,

$$\Phi(\varepsilon, \sigma) U_\alpha = \begin{cases} U_{\sigma(\alpha)} & \text{if } \sigma(0) = 0 \\ U_{\sigma(\alpha)} - U_{\sigma(0)} & \text{if } \sigma(0) \neq 0, \quad \sigma(\alpha) \neq 0 \\ -U_{\sigma(0)} & \text{if } \sigma(0) \neq 0, \quad \sigma(\alpha) = 0 \end{cases}$$

for all $i = 0,1,\ldots,n$ and $\alpha = 1,\ldots,n$.
Then we can prove that $\Phi$ is an isomorphism of $(Z_2)^{n+1} \times S_{n+1}$ onto the subgroup of $GL(\underline{g}_n)$ consisting of all isometric automorphisms of $\underline{g}_n$. Further, the symmetries of $(G_n, g)$ at the identity e correspond to the pairs $(\varepsilon, \sigma) \in (Z_2)^{n+1} \times S_{n+1}$ for which $\varepsilon = (\varepsilon_0, \ldots, \varepsilon_n)$ contains an odd number of $(-1)$'s and $\sigma$ is a full cycle. Hence we see that all symmetries of $(G_n, g)$ at e have order $2n+2$ and they are conjugate to each other.

The irreducibility of $(G_n, g)$ is easily proved from the fact that the Lie algebra $\underline{g}_n$ is not a direct sum of two proper ideals. As a conclusion, the space $(G_{\frac{1}{2}(m-2)}, g)$ satisfies the requirement of Theorem 1.

B) The affine case.

We shall start from the same matrix group $G_n$ as before, $G_n \approx R^{2n+1}(x_0, x_1, \ldots, x_n, u_1, \ldots, u_n)$. Let $H_n$ denote the 1-dimensional subgroup of $G_n$ given by $x_0 = x_1 = \ldots = x_{n-1} = -x_n = t$, $u_1 = \ldots$ $\ldots = u_n = 0$, and consider the homogeneous space $M_n = G_n/H_n$. It is easy to see that $M_n \approx R^{2n}$ for all $n \geq 2$. The Lie algebra $\underline{g}_n$ possesses a basis $\{Y_0, Y_1, \ldots, Y_n, U_1, \ldots, U_n\}$ with the multiplication table given by

$$\left| \begin{array}{l} [Y_i, Y_j] = [U_\alpha, U_\beta] = 0, \quad [Y_0, U_\alpha] = Y_\alpha , \\ [Y_\alpha, U_\beta] = \frac{1 + \delta_{\alpha\beta}}{n} \left( \sum_{p=0}^{n} Y_p \right) - \delta_{\alpha\beta} Y_\alpha , \\ \quad i,j = 0,1,\ldots,n; \quad \alpha, \beta = 1,\ldots,n. \end{array} \right| \qquad (1)$$

Here $\underline{h}_n$ is spanned by $Y_0$. Put $\underline{m}_n = (Y_1, \ldots, Y_n, U_1, \ldots, U_n)$. $G_n/H_n$ is reductive with respect to the decomposition $\underline{g}_n = \underline{h}_n + \underline{m}_n$. Let $\tilde{\nabla}_n$ be the canonical connection. By a routine calculation we find that the Lie algebra $\underline{a}_n$ of the full affine group $A(M_n, \tilde{\nabla}_n)$ possesses a basis $\{Y_0, Y_1, \ldots, Y_n, U_1, \ldots, U_n, U\}$, for which the multiplication is given by the relations (1) and by the additional relations $[U,Y_i] = Y_i$, $[U,U_\alpha] = 0$. In particular, $\underline{g}_n$ is a subalgebra of codimension 1 of $\underline{a}_n$. $\underline{a}_n$ is obviously solvable and hence the identity component of $A(M_n,\tilde{\nabla}_n)$ is solvable. Here $G_n$ is the transvection group of $(M_n,\tilde{\nabla}_n)$ and $\underline{g}_n$ is the transvection algebra.

Now there is a (1-1)-correspondence between the set of all admissible s-structures on $(M_n,\tilde{\nabla}_n)$ and the set of all "admissible" automorphisms $\Phi$ of $\underline{g}_n$ characterized by the properties $\Phi(\underline{m}_n) = \underline{m}_n$, $(\underline{g}_n)^{\Phi} = \underline{h}_n$.

Further, consider the set $C_{n+1}$ of all full cycles among the permutations of the index set $\{0,1,\ldots,n\}$. To any pair $(r,\sigma) \in R\times C_{n+1}$ we consider a linear transformation $F(r,\sigma)$ of $\underline{g}_n$ given by

$$F(r,\sigma)(Y_o) = Y_o,$$

$$F(r,\sigma)(Y_\alpha) = \begin{cases} Y_{\sigma(\alpha)} - Y_{\sigma(0)} & \text{if } \sigma(\alpha) \neq 0, \\ -Y_{\sigma(0)} & \text{if } \sigma(\alpha) = 0, \end{cases} \qquad (2)$$

$$F(r,\sigma)(U_\alpha) = \begin{cases} r(Y_{\sigma(\alpha)} - Y_{\sigma(0)}) + U_{\sigma(\alpha)} - U_{\sigma(0)} & \text{if } \sigma(\alpha) \neq 0, \\ -rY_{\sigma(0)} - U_{\sigma(0)} & \text{if } \sigma(\alpha) = 0. \end{cases}$$

It is not difficult to show that $F$ is a bijective map of $R\times C_{n+1}$ onto the set of all admissible automorphisms of $\underline{g}_n$. Hence each admissible automorphism of $\underline{g}_n$ is either of infinite order, or of order $n+1$, and the same holds for admissible s-structures on $(M_n,\tilde{\nabla}_n)$.

Because the Lie algebra $\underline{g}_n$ is not a direct sum of its ideals, the transvection group $G_n = Tr(M_n,\tilde{\nabla}_n)$ is not a direct product, and hence our affine reductive space $(M_n,\tilde{\nabla}_n)$ cannot be a direct product. (Cf. Theorem I.35.) In particular, $(M_n,\tilde{\nabla}_n)$ is primitive as a generalized affine symmetric space.

We only have to put $m = n + 1$ and Theorem 2 is proved.

## N o t e 2. Irreducible generalized affine symmetric spaces.

An affine reductive space $(M,\tilde{\nabla})$ is said to be <u>irreducible</u> if, for any $p\in M$, the restricted holonomy group with reference point $p$ acts irreducibly on the tangent space $T_p(M)$.

<u>Theorem 1</u>. If a g.a.s. space $(M,\tilde{\nabla})$ is irreducible, then its transvection group $Tr(M)$ is semi-simple. Moreover, for each admissible regular s-structure $\{s_x\}$ on $(M,\tilde{\nabla})$, the identity component of the automorphism group $Aut(M,\{s_x\})$ coincides with $Tr(M)$.

Proof. Choose an origin $o \in M$. Let $\{s_x\}$ be an admissible regular s-structure on $(M,\tilde{\nabla})$, $\mathrm{Aut}(M) = \mathrm{Aut}(M,\{s_x\})$ the full automorphism group and $\sigma : \mathrm{Aut}(M) \longrightarrow \mathrm{Aut}(M)$ the Lie group automorphism corresponding to the symmetry $s_o$. Let $G$ be a connected $\sigma$-invariant subgroup of $\mathrm{Aut}(M)$ acting transitively on $M$. Then $\mathrm{Tr}(M) \subseteq G$ according to II.39, and $(G, G_o, \sigma|_G)$ is a regular homogeneous s-manifold. (Cf. II.38.) We shall write the last triplet in the brief form $(G,H,\sigma)$.

According to I.26, the linear isotropy representation of $(\mathrm{Tr}(M))_o$ in the tangent space $T_o(M)$ is faithful and the image can be identified with the holonomy group of $(M,\tilde{\nabla})$ with reference point $o$. Now the identity component $H^{\bullet}$ of $H$ contains the identity component of $(\mathrm{Tr}(M))_o$ and thus it acts irreducibly on the tangent space $T_o(M)$. Let $\underline{g} = \underline{m} + \underline{h}$ be the canonical decomposition making $G/H$ a reductive homogeneous space; $\mathrm{ad}(\underline{h})(\underline{m}) \subset \underline{m}$. Then the irreducibility of $H^{\bullet}$ can be reformulated as follows: the restriction of the Lie algebra $\mathrm{ad}_{\underline{g}}(\underline{h})$ to the subspace $\underline{m} \subset \underline{g}$ is irreducible on $\underline{m}$. The differential $\sigma_*$ of $\sigma$ is a Lie algebra automorphism $\sigma_* : \underline{g} \longrightarrow \underline{g}$ such that $\sigma_*|_{\underline{h}} = \mathrm{id}$, and $\underline{g} = \underline{m} + \underline{h}$ is the Fitting decomposition with respect to the linear endomorphism $A = \sigma_* - \mathrm{Id}$, i.e., $\underline{h} = \underline{g}_{0A}$, $\underline{m} = \underline{g}_{1A}$. Recall that the Lie algebra of $(\mathrm{Tr}(M))_o$ is $[\underline{m},\underline{m}] \cap \underline{h}$, and that it is isomorphic to the holonomy algebra with reference point $o$. (Cf. proof of I.30.)

Denote by $\underline{r}$ the radical of $\underline{g}$, and suppose $\underline{r} \neq 0$. Then we can find the maximum integer $k \geq 0$ such that $\mathcal{D}^k\underline{r} \neq 0$, where $\{\mathcal{D}^i\underline{r}\}$ denotes the derived series of $\underline{r}$. Now, $\mathcal{D}^k\underline{r}$ is invariant with respect to the automorphism $\sigma_*$ and also with respect to $A$. Using the Fitting decomposition of $\mathcal{D}^k\underline{r}$ with respect to $A$ we get easily

$$\mathcal{D}^k\underline{r} = \mathcal{D}^k\underline{r} \cap \underline{h} + \mathcal{D}^k\underline{r} \cap \underline{m}.$$

Because $\mathcal{D}^k\underline{r}$ is an ideal in $\underline{g}$ and $\mathrm{ad}(\underline{h})\underline{m} \subset \underline{m}$, we get $\mathrm{ad}(\underline{h})(\mathcal{D}^k\underline{r} \cap \underline{m}) \subset \mathcal{D}^k\underline{r} \cap \underline{m}$. From the irreducibility of $\mathrm{ad}(\underline{h})$ on $\underline{m}$ we see that either $\mathcal{D}^k\underline{r} \cap \underline{m} = \underline{m}$ or $\mathcal{D}^k\underline{r} \cap \underline{m} = 0$.

The first condition implies $\underline{m} \subset \mathcal{D}^k\underline{r}$ and $[\underline{m},\underline{m}] \subset [\mathcal{D}^k\underline{r}, \mathcal{D}^k\underline{r}] = \mathcal{D}^{k+1}\underline{r} = 0$. In particular, we get $[\underline{m},\underline{m}] \cap \underline{h} = 0$; thus the holonomy algebra is trivial - a contradiction. In the second case we get $\mathcal{D}^k\underline{r} \subset \underline{h}$ and $[\mathcal{D}^k\underline{r},\underline{m}] \subset [\underline{h},\underline{m}] \subset \underline{m}$, $[\mathcal{D}^k\underline{r},\underline{m}] \subset [\mathcal{D}^k\underline{r},\underline{g}] \subset \mathcal{D}^k\underline{r}$. Thus $[\mathcal{D}^k\underline{r},\underline{m}] = 0$ and hence $\mathcal{D}^k\underline{r} = 0$ because $\mathrm{ad}(\underline{h})$ is effective on $\underline{m}$. (Recall that the linear isotropy representation of $H^{\bullet}$ in $T_o(M)$ is faithful because $H^{\bullet}$ consists of affine transformations.) This also contradicts to the assumption $\underline{r} \neq 0$. Hence $\underline{g}$ is semi-simple.

It remains to show that $\underline{g} = \underline{m} + [\underline{m},\underline{m}]$ = the Lie algebra of $Tr(M)$. Let $\Phi$ denote the Killing form of $\underline{g}$; then $\sigma_*$ is an isometry with respect to the (indefinite) scalar product $\Phi$. We have $\Phi(\underline{m},\underline{h}) = 0$. In fact, choose $X \in \underline{m}$, $U \in \underline{h}$; then because $A = \sigma_* - Id$ is a linear automorphism on $\underline{m}$, we have $X = AY$ for some $Y \in \underline{m}$. Hence $\Phi(X,U) = \Phi(\sigma_* Y - Y, U) = \Phi(\sigma_* Y, \sigma_* U) - \Phi(Y,U) = 0$. Now, a basic property of a Killing form is that the orthogonal complement of an ideal is also an ideal. In particular, the ideal $(\underline{m} + [\underline{m},\underline{m}])^{\perp}$ is contained in $\underline{m}^{\perp} = \underline{h}$, and hence $(\underline{m} + [\underline{m},\underline{m}])^{\perp} = 0$ because $G$ is acting effectively on $G/H = M$. (Cf. [F3].) □

Remark. Consider the opposite situation where the restricted holonomy group of $(M,\tilde{\nabla})$ is trivial. Then the subalgebra $[\underline{m},\underline{m}] \cap \underline{h}$ is trivial and $[\underline{m},\underline{m}] \subset \underline{m}$. Hence $\tilde{R} = 0$, and the group $Tr(M)$ is solvable according to II.41.

* Theorem 2. An irreducible g.a.s. space $(M,\tilde{\nabla})$ is either

a) locally symmetric, or

b) 3-symmetric but not k-symmetric for any $k \neq 3$.

Proof. Let $\{s_x\}$ be an admissible regular s-structure on $(M,\tilde{\nabla})$ such that $(s_x)^3 \neq Id$. Let $(V,S,\tilde{R},\tilde{T})$ be the infinitesimal model of $(M,\{s_x\})$ and $J$ the set of all eigenvalues of $\{s_x\}$. From the proof of Theorem IV.40 we see the following: if $\lambda \in J$ is real, then the eigenspace $V_\lambda \subset V$ is invariant with respect to the automorphism group of $(V,S,\tilde{R},\tilde{T})$. If $\lambda \in J$ is imaginary, then the real subspace $W_\alpha = (V_\alpha \oplus V_{\bar{\alpha}}) \cap V$ is invariant with respect to the same group.

Let us identify $V$ with a tangent space $M_p$; then the invariance with respect to the automorphisms of $(V,S,\tilde{R},\tilde{T})$ implies the invariance with respect to the holonomy group of $(M,\tilde{\nabla})$ with reference point p. Because $(M,\tilde{\nabla})$ is irreducible, $V$ cannot contain a proper invariant subspace $W$. Hence the only possibilities are the following: (i) $J = \{\lambda\}$, $\lambda$ real, $\lambda^2 \neq \lambda$.

(ii) $J = \{\alpha, \bar{\alpha}\}$, $\alpha$ imaginary. Here we have $\alpha^3 \neq 1$ and hence $\alpha^2 \neq \alpha, \bar{\alpha}$.

From the first part of Proposition IV.38 we get, in each case, $\tilde{T} = 0$. Thus $(M,\tilde{\nabla})$ is locally symmetric. □

---

References: [F3],[KN II].

## N o t e 3. Generalized pointwise symmetric spaces.

Definition. A generalized pointwise symmetric Riemannian space (or briefly, a GPS space) is a Riemannian manifold admitting a (not necessarily regular) Riemannian s-structure. (Cf. 0.4.) Order of a GPS space is the minimum of orders of its admissible Riemannian s-structures.

Reformulations of 0.1. and 0.6. give the following:

Theorem 1. Every GPS space is homogeneous.

Theorem 2. Every GPS space is of finite order.

The proofs of the following theorems are easy modifications of the proofs for the "regular" case:

Theorem 3. Let $(M,g)$ be a simply connected GPS space, and $M = M_o \times M_1 \times \dots \times M_r$ its de Rham decomposition. Then each factor $M_i$ is a GPS space.

Theorem 4. For each $k \geq 2$ there is a compact GPS space $(M,g)$ of order $k$ such that the group $I(M,g)^{\bullet}$ is semi-simple.

-------

Let us remark that GPS spaces of order 2, as well as those of dimension 2, are Riemannian symmetric spaces. The purpose of this Note is to present a family of GPS spaces which are not generalized symmetric. (See [K4].)

Consider the complex Cartesian space $C^{2n+1}[z^1,\dots,z^{2n+1}]$ endowed with a Hermitian metric

$$g_{\lambda} = \sum_{i=1}^{2n+1} dz^i d\bar{z}^i + \lambda\Big(\sum_{i=1}^{2n+1} z^i d\bar{z}^i\Big)\Big(\sum_{j=1}^{2n+1} \bar{z}^j dz^j\Big), \qquad \lambda \neq 0, \quad \lambda > -1.$$

Let us consider the sphere $S^{4n+1}$ defined by $\sum_{i=1}^{2n+1} z^i\bar{z}^i = 1$, and the real Riemannian metric $\hat{g}_{\lambda}$ on $S^{4n+1}$ induced by $g_{\lambda}$. (Here the real coordinates are introduced putting $z^j = x^j + iy^j$, $j = 1,\dots,2n+1$.)

Let us remark that, for $-1 < \lambda < 0$, the manifold $(S^{4n+1},\hat{g}_{\lambda})$ is homothetic to a quadric of the complex projective space $P_{2n+1}(C)$

endowed with the natural Fubini-Study metric (cf. [KN II], p.160).

For n=1, we obtain the Type 6a) of the classification list of the generalized symmetric Riemannian spaces of dimension 5 (Chapter VI).

Now, we are going to prove

**Theorem 5.** For $n \geq 2$, the Riemannian manifold $(S^{4n+1}, \hat{g}_\lambda)$ is generalized pointwise symmetric of order 4 but it is not generalized symmetric.

**Proof.** We shall prove the first part of our Theorem. Let us define the origin $o$ of $S^{4n+1}$ by $o = (0,\dots,0,1) \in C^{2n+1}$. The transformation of $C^{2n+1}$ given by $(z^{2i-1})' = -\bar{z}^{2i}$, $(z^{2i})' = \bar{z}^{2i-1}$ $(i = 1,\dots,n)$, $(z^{2n+1})' = \bar{z}^{2n+1}$, induces a transformation $\tilde{s}_o$ of $S^{4n+1}$ with a fixed point $o$. Clearly, $\tilde{s}_o$ is an isometry of $(S^{4n+1}, \hat{g}_\lambda)$. We can see easily that the tangent map $(\tilde{s}_o)_{*o}$ has no non-zero fixed vectors in the tangent space $(S^{4n+1})_o$, and hence $o$ is an isolated fixed point of $\tilde{s}_o$. Moreover, we have $(\tilde{s}_o)^4$ = identity.

The group $U(2n+1)$ of all unitary transformations of $C^{2n+1}$ (with respect to its natural structure of a linear Hermitian space) preserves the metric $g_\lambda$ and it acts transitively and effectively on $S^{4n+1}$. Thus $U(2n+1)$ can be considered as a group of isometries of the Riemannian manifold $(S^{4n+1}, \hat{g}_\lambda)$. Let $U(2n)$ denote the isotropy group of $U(2n+1)$ at $o$; and let $\pi: U(2n+1) \longrightarrow U(2n+1)/U(2n) \cong S^{4n+1}$ be the bundle projection. Choose a (not necessarily continuous!) cross-section $\varphi: S^{4n+1} \longrightarrow U(2n+1)$. Then a Riemannian s-structure of order 4 on $(S^{4n+1}, \hat{g}_\lambda)$ can be defined by the formula $\tilde{s}_x = \varphi(x) \circ \tilde{s}_o \circ \varphi(x)^{-1}$, $x \in S^{4n+1}$.

Let us remark that $(S^{4n+1}, \hat{g}_\lambda)$ is not locally symmetric (an easy calculation) and that it is of odd dimension. Thus, the order of the space cannot be 2 or 3, and it is equal to 4.

We shall now prove the second part of the Theorem, which is non-trivial.

In the following, $SO(4n+2)$, $U(2n+1)$ and $SU(2n+1)$ will always denote the transformation groups of $S^{4n+1}$ which are induced by the corresponding transformation groups of the given real space $R^{4n+2}$ and of the complex space $C^{2n+1}$.

**Lemma.** Let $K$ be a connected Lie group of isometries of $(S^{4n+1}, \hat{g}_\lambda)$ acting transitively on $S^{4n+1}$. Then $K \supset SU(2n+1)$.

Proof. According to Montgomery-Samelson [MS], and Borel [B1],[B2] (see the references at the end of this Note), each compact connected Lie transformation group acting transitively on $S^{4n+1}$ is isomorphic to one of the following groups: $SO(4n+2)$, $U(2n+1)$, $SU(2n+1)$. Let $G$ be the identity component of the full isometry group $I(S^{4n+1},\hat{g}_\lambda)$, then $G \supset U(2n+1)$. $G$ cannot be isomorphic to $SO(4n+2)$; otherwise $\hat{g}_\lambda$ would be a metric of constant curvature. Thus $G = U(2n+1)$.

Let $K$ be an arbitrary connected and transitive group of isometries of $(S^{4n+1},\hat{g}_\lambda)$; then $K \subset U(2n+1)$. If $K$ is isomorphic to $U(2n+1)$, then $K = U(2n+1)$ and Lemma is proved. Let now $K$ be isomorphic to $SU(2n+1)$. Then the Lie algebra $\underline{k}$ is isomorphic to $\underline{su}(2n+1)$, and $\underline{k} \subset \underline{u}(2n+1)$. On the other hand, we have $\underline{u}(2n+1) = \underline{su}(2n+1) \oplus R$ (direct sum), and the subalgebra $\underline{su}(2n+1)$ is simple. Hence it follows $\underline{k} = \underline{su}(2n+1)$, and consequently, $K = SU(2n+1)$. □

Suppose now that there is a Riemannian regular s-structure $\{s_x\}$ on $(S^{4n+1},\hat{g}_\lambda)$, and denote $K = Tr(S^{4n+1},\{s_x\}) \subset I(S^{4n+1},\hat{g}_\lambda)$. According to our Lemma, $K \supset SU(2n+1)$. For the isotropy group $K_o$ at $o$ we have $K_o \supset SU(2n)$ (= the subgroup of $SU(2n+1)$ leaving all the points $(0,\dots,0,e^{i\varphi})$ of $S^{4n+1}$ fixed). The transformation $s_o$ commutes with all elements of $K_o$ and, in particular, it commutes with all elements of $SU(2n)$.

Consider the tangent space $(S^{4n+1})_o$. It is generated by the vectors $e_i = (\frac{\partial}{\partial x^i})_o$, $f_j = (\frac{\partial}{\partial y^j})_o$, where $i = 1,\dots,2n$, $j = 1,\dots,2n+1$. Here $f_{2n+1}$ is orthogonal to the $4n$-dimensional subspace $V$ generated by $e_i$, $f_i$ for $i = 1,\dots,2n$.

Let $H$ denote the image of the real isotropy representation of $SU(2n)$ in the tangent space $(S^{4n+1})_o$, and $S_o = (s_o)_{*o}$. All linear transformations $h \in H$, and also $S_o$, are orthogonal transformations of $(S^{4n+1})_o$ with respect to the scalar product $(\hat{g}_\lambda)_o$. $H$ acts transitively on the subspace $V$, and all fixed vectors with respect to $H$ are of the form $\lambda \cdot f_{2n+1}$. $S_o$ commutes with each $h \in H$ and hence $S_o(f_{2n+1})$ is a fixed vector with respect to $H$. Thus $S_o(f_{2n+1}) = \pm f_{2n+1}$, and since $S_o$ does not admit non-zero fixed vectors, $S_o(f_{2n+1}) = -f_{2n+1}$. Also, the subspace $V$ is invariant with respect to $S_o$.

Let $\underline{h}$ denote the Lie algebra of $H$. For every pair $(r,s)$, $1 \leq r < s \leq 2n$, consider the endomorphisms $B_{rs}$, $C_{rs} \in \underline{h}$ defined by

$$B_{rs}(e_r) = e_s, \quad B_{rs}(f_r) = f_s, \quad B_{rs}(e_s) = -e_r, \quad B_{rs}(f_s) = -f_r,$$

$$C_{rs}(e_r) = -f_s, \quad C_{rs}(f_r) = e_s, \quad C_{rs}(e_s) = -f_r, \quad C_{rs}(f_s) = e_r,$$

$B_{rs}(e_i) = B_{rs}(f_i) = C_{rs}(e_i) = C_{rs}(f_i) = 0$ for $i \neq r,s$.

Let $S_o$ satisfy

$$S_o(e_i) = \sum_{j=1}^{2n}[a_i^j e_j + b_i^j f_j], \quad S_o(f_i) = \sum_{j=1}^{2n}[c_i^j e_j + d_i^j f_j], \quad i=1,\dots,2n.$$

From the relations
$$(B_{rs}\circ S_o)(e_i) = (S_o\circ B_{rs})(e_i), \qquad i \neq r,s$$
$$(B_{rs}\circ S_o)(f_i) = (S_o\circ B_{rs})(f_i)$$
we get $a_i^j = b_i^j = c_i^j = d_i^j = 0$, for all $i$, $j$ such that $1 \leq i \neq j$. (For this step, the inequality $n \geq 2$ is decisive.) From the relations

$$(B_{rs}\circ S_o)(e_r) = (S_o\circ B_{rs})(e_r), \qquad (B_{rs}\circ S_o)(f_r) = (S_o\circ B_{rs})(f_r)$$

we get $a_r^r = a_s^s$, $b_r^r = b_s^s$, $c_r^r = c_s^s$, $d_r^r = d_s^s$, $1 \leq r < s \leq 2n$.
Finally, from the relation $(C_{rs}\circ S_o)(e_r) = (S_o\circ C_{rs})(e_r)$ we get $a_r^r = d_s^s = a$, $b_r^r = -c_s^s = b$, $1 \leq r < s \leq 2n$.
We have obtained $S_o(e_j) = ae_j + bf_j$, $S_o(f_j) = -be_j + af_j$ for $1 \leq j \leq 2n$, where $a^2 + b^2 = 1$, and $S_o(f_{2n+1}) = -f_{2n+1}$.
Or, in the complex form, $S_o((\frac{\partial}{\partial z^j})_o) = e^{i\varphi}((\frac{\partial}{\partial z^j})_o)$, $j = 1,\dots,2n$
$S_o(f_{2n+1}) = -f_{2n+1}$.

Now, let us denote by $Z_1,\dots,Z_{2n+1}$ the complex vector fields on $S^{4n+1}$ which are tangent components of the vector fields $\frac{\partial}{\partial z^1},\dots$ $\dots,\frac{\partial}{\partial z^{2n+1}}$ respectively. Let $\nabla$, $R$ denote the Riemannian connection and the curvature tensor field of the metric $\hat{g}_\lambda$ respectively. After a lenghty but routine calculation we obtain

$(\nabla_{Z_2}R)_o(Z_1,\bar{Z}_1,Z_{2n+1},\bar{Z}_2) \neq 0$, i.e.
$(\nabla_{\frac{\partial}{\partial z^2}}R)_o((\frac{\partial}{\partial z^1})_o,(\frac{\partial}{\partial \bar{z}^1})_o, f_{2n+1},(\frac{\partial}{\partial \bar{z}^2})_o) \neq 0$. If we apply the map $S_o$ to each argument, and use the invariance of $(\nabla R)_o$ with respect to $S_o$, we obtain that the previous term is zero - a contradiction. □ □

Problem 1. Let us consider the following classes of Riemannian manifolds:

- $S_1$ ... the class of all GPS spaces,
- $S_2$ ... the class of all Riemannian manifolds admitting a continuous s-structure,
- $S_3$ ... the class of all Riemannian manifolds admitting a smooth s-structure,

$S_4$ ... the class of generalized symmetric Riemannian spaces. Is each $S_{i+1}$ a proper subclass of $S_i$, $i = 1,2,3$ ?

Problem 2. Let $(M,\nabla)$ be an affine manifold admitting an affine s-structure $\{s_x\}$. (Here each mapping $s_x$ is an affine transformation with an isolated fixed point $x$, and we do not make any other assumptions.) In the connection with Theorem 1 (F.Brickel) decide:

(a) Is $(M,\nabla)$ homogeneous ? (b) Is $(M,\nabla)$ complete ?

---


Special references to this Note:

[B1] A.Borel: Some remarks about Lie groups transitive on spheres and tori. Bull. Amer. Math. Soc. 55 (1948), 580-586.

[B2] A.Borel: Le plan projectif des octaves et les sphères comme espaces homogènes, C.R. Acad. Sci. Paris 230 (1950), 1378-1380.

[MS] D.Montgomery, H.Samelson: Transformation groups on spheres. Ann. of Math. 44 (1943), 454-470.

Other references: [GL2],[K1],[K3],[K4],[K5],[LO].


## N o t e 4. Non-parallel s-structures on symmetric spaces.

A regular s-structure $\{s_x\}$ on a Riemannian manifold $(M,g)$ was said to be parallel or non-parallel in accordance with the relation $\nabla S = 0$ or $\nabla S \neq 0$ (see Chapter 0). If $(M,g)$ is a Riemannian symmetric space, then the regular s-structure of geodesic symmetries of $(M,g)$ induces the tensor field $S = -I$ and thus it is parallel. Now, we ask whether there exist also non-parallel regular s-structures on Riemannian symmetric spaces. In particular, what exactly are the Riemannian symmetric spaces having such a degree of freedom ?

In connection with the classification of generalized symmetric Riemannian spaces of dimension $n \leq 5$, we shall be able to solve our problem (under the usual restrictions) for dimension $n \leq 5$. We start with some preliminary remarks.

Firstly, for each Riemannian symmetric space $(M,g)$ with a non--parallel regular s-structure, the universal covering manifold $(\tilde{M},\tilde{g})$ is symmetric and it also bears a non-parallel regular s-structure. Thus we can limit ourselves to the simply connected spaces.

Secondly, let $(M,g,\{s_x\})$ be a simply connected Riemannian regular s-manifold, where $(M,g)$ is symmetric and $\{s_x\}$ is non-parallel. Suppose this s-manifold be reducible, i.e., $(M,g,\{s_x\}) =$

$= (M_1, g_1, \{s_u^1\}) \times (M_2, g_2, \{s_v^2\})$. Then $(M_1, g_1)$, $(M_2, g_2)$ are both symmetric and at least one of the s-structures $\{s_u^1\}$, $\{s_v^2\}$ is non-parallel. Hence reducible Riemannian regular s-manifolds and reducible infinitesimal s-manifolds are not essential for our problem. (On the other hand, we must not omit reducible Riemannian manifolds from our considerations, as we shall see below.)

Finally, if a simply connected space $(M,g)$ admits a non-parallel (and irreducible) regular s-structure $\{s_x\}$, then it also admits a non-parallel regular s-structure $\{s'_x\}$ with a system of eigenvalues $(\theta'_i) \in \mathfrak{D}^{*n}$.

Obviously, a Riemannian regular s-structure $\{s_x\}$ on $(M,g)$ is non-parallel if and only if the torsion tensor $\tilde{T} \neq 0$. In fact, according to Formula (14), Chapter III, we have $\tilde{T} = 0 \Longleftrightarrow \tilde{\nabla} = \nabla$. Now, according to Theorem 0.33, $\tilde{\nabla} = \nabla$ holds if and only if $\nabla S = 0$.

Let us resume our classification procedure used in Chapter VI. The main steps of the procedure (for a fixed dimension n) were the following:

a) The classification of all (non-isomorphic) irreducible Riemannian infinitesimal s-manifolds with the torsion tensor $\tilde{T} \neq 0$ and such that their systems of eigenvalues belong to $\mathfrak{D}^{*n}$.

b) The construction of irreducible, simply connected Riemannian regular s-manifolds $(M, g, \{s_x\})$ from the given infinitesimal s-manifolds.

c) Making the list of all non-isometric generalized symmetric Riemannian spaces $(M,g)$ that are obtained in this way.

d) The omission of all spaces $(M,g)$ in the list which proved to be locally symmetric.

e) The omission of all spaces $(M,g)$ in the list which proved to be reducible as Riemannian manifolds.

Then it is clear that the spaces of interest are exactly those which were omitted in the step d).

**Theorem 1.** The symmetric spaces of dimensions 2, 3 and 4 (both simply connected or not) admit only parallel regular s-structures.

**Proof.** In the dimension $n = 2$, the torsion tensor $\tilde{T}$ is always zero. Now, let us review the classification procedure for the dimensions $n = 3, 4$ (see Chapter VI). We see that all generalized symmetric Riemannian spaces obtained by this procedure proved to be locally non-symmetric.

As for the dimension $n = 5$, the details of the classification were not given here, and the reader is advised to see the original booklet [K3]. Very briefly, the symmetric spaces occur here for some "singular" values of parameters, namely in the following cases:

a) For $\lambda + \bar{\lambda} = 0$, $\alpha = 0$, $c = 0$ in Type 4),
b) for $\mu = 0$ in Type 6a),
c) for $\mu = 1$ in Type 8a).

We summarize with some more details:

Theorem 2. The only simply connected Riemannian symmetric spaces of dimension 5 admitting non-parallel regular s-structures are $E^5$, $S^5(r)$, and $E^3 \times S^2(r)$.

More specifically:

[a] On the space $E^5$, we obtain non-parallel s-structures $\{s_x^{\varrho}\}$ of order 4, depending on a real parameter $\varrho > 0$ as follows: we identify $E^5$ with the space $C^2(z,w) \times R^1(t)$ and define a symmetry $\sigma_o$ at the origin $o = (0,0;\ 0)$ by the relations $z' = iw$, $w' = iz$, $t' = -t$. Further, consider for each $\varrho > 0$ the simply connected transitive transformation group

$$G_{\varrho}:\ z' = e^{\varrho i t_o} z + z_o, \quad w' = e^{-\varrho i t_o} w + w_o, \quad t' = t + t_o.$$

Then the set $\{s_x^{\varrho}:\ x \in E^5\}$ coincides with the set $\{g \circ \sigma_o \circ g^{-1}:\ g \in G_{\varrho}\}$.

[b] On the space $S^5(r)$, we obtain a non-parallel s-structure $\{s_x\}$ of order 4 as follows: we identify $S^5(r)$ with the submanifold $z^1\bar{z}^1 + z^2\bar{z}^2 + z^3\bar{z}^3 = r^2$ of the complex euclidean space $C^3(z^1,z^2,z^3)$ and define a symmetry $\sigma_o$ at the "origin" $o = (0,0,r) \in C^3$ of $S^5$ by the relations $(z^1)' = \bar{z}^2$, $(z^2)' = -\bar{z}^1$, $(z^3)' = \bar{z}^3$. Then the set $\{s_x:\ x \in S^5(r)\}$ coincides with the set $\{g \circ \sigma_o \circ g^{-1}:\ g \in SU(3)\}$.

[c] On the space $E^3 \times S^2(r)$, we obtain a non-parallel s-structure $\{s_x\}$ of order 4 as follows: we identify $E^3 \times S^2(r)$ with a sphere bundle over the base space $E^3(x,y,z)$, namely with the set of all pairs $(m,t)$, where $m \in E^3$, $t \in T_m(E^3)$, $|t| = r$. Let $I^e(E^3)$ be the group of all orientation-preserving euclidean motions of $E^3$ and let $J^1(I^e(E^3))$ denote its first prolongation group acting on the sphere bundle $E^3 \times S^2(r)$. We can identify the tangent bundle $T(E^3)$ with a space $E^6(x,y,z;\alpha,\beta,\gamma)$ in a natural way, and the space

$E^3 \times S^2(r)$ with the submanifold of $E^6$ given by the relation $\alpha^2 + \beta^2 + \gamma^2 = r^2$. Finally, consider the transformation of $E^6$ given by the relations $x' = -y$, $y' = x$, $z' = -z$, $\alpha' = \beta$, $\beta' = -\alpha$, $\gamma' = \gamma$. The induced transformation $\sigma_o$ of $E^3 \times S^2(r)$ is then a symmetry of order 4 of this submanifold at the point $(0,0,0;\ 0,0,r)$. Now, the wanted regular s-structure $\{s_x : x \in E^3 \times S^2(r)\}$ coincides with the set $\{g \circ \sigma_o \circ g^{-1} : g \in J^1(I^e(E^3))\}$.

Reference: [K3], Chapter 16.

## N o t e 5. Some advanced results.

J.A.Wolf and A.Gray, [WoG], have studied the structure and geometry of homogeneous spaces $G/K$, where $G$ is a reductive Lie group and $K$ is an open subgroup of the fixed point set $G^\theta$ of a semi-simple automorphism $\theta$ of $G$. (Recall that a Lie group $G$ is said to be reductive if the adjoint group $\mathrm{ad}\, G$ is semi-simple.) Here a complete classification is given of all spaces $G/K$ which are simply connected and come from an automorphism $\theta$ of order 3. For the automorphisms $\theta$ of arbitrary finite order only a general method of classification is presented. (*).

Geometric applications of the previous results are given in a subsequent paper by A.Gray ([Gr]). The author presents a complete classification of all simply connected pseudo-Riemannian "3-symmetric spaces" (in our terminology: pseudo-Riemannian regular s-manifolds of order 3) such that the group $\mathrm{Aut}(M) \cap I(M)$ of isometric automorphisms is a reductive Lie group. Essentially, this gives the classification of all 3-symmetric pseudo-Riemannian spaces of the semi-simple type.

The stress is given on the study of invariant almost complex and almost Hermitian structures on the corresponding regular s-manifolds (in the sense of III.46, III.47, III.49), in particular, on the study of nearly Kähler structures. (A nearly Kähler structure $J$ is characterized by the identity $(\nabla_X J)(X) = 0$ for all vector fields $X$, where $\nabla$ denotes the pseudo-Riemannian connection.) The following is proved: Each pseudo-metrizable regular s-manifold of order 3 and "of semi-simple type" can be equipped with such a pseudo-Riemannian met-

(*) A complete description (rather than an explicit classification) of all finite order automorphisms on semi-simple Lie algebras has been found by Viktor Kac (cf. [Ka] and [H*]).

ric $g$ that the corresponding invariant almost Hermitian structure $J$ is nearly Kählerian. If the manifold is indecomposable, then such a metric is unique up to a scalar multiple. Conversely, if $M$ is an analytic pseudo-Riemannian manifold with a nearly Kähler structure $J$ satisfying the additional condition $(\nabla_X R)(X,JX,X,JX) = 0$, then $M$ can be made a (pseudo-Riemannian) locally regular s-manifold of order 3 for which the complex structure $J$ is invariant by the symmetries. In particular, $M$ is locally 3-symmetric.

Using the methods of [WoG], A.S.Fedenko has classified all homogeneous regular s-manifolds $(G,H,\sigma)$ such that $G$ is a simple Lie group (a classical or exceptional one) and $\sigma$ is a periodic automorphism ([Fe3]).

Making use of the same methods, A.J.Ledger and R.B.Pettit have classified all compact "quadratic" s-manifolds ([LP1]). (A quadratic s-manifold is, by definition, a metrizable regular s-manifold $(M,\{s_x\})$ for which the symmetry tensor field $S$ has a quadratic minimal polynomial, or equivalently, for which $S$ possesses exactly two distinct eigenvalues, which are complex conjugate.)

In the paper [LP2], the same authors give an example of a 3-symmetric Riemannian space which is not homeomorphic to any ordinary symmetric space. The corresponding underlying manifold is $M = SU(3)/T^2$, where $T^2$ is the maximal torus of $SU(3)$.

In the classification list by A.Gray, as well as in the other classification results mentioned in this place, the spaces of solvable of "mixed" type are not involved. <u>For the comparison</u>, it may be interesting to go through the present situation in the theory of the pseudo-Riemannian symmetric spaces: The complete classification of those of semi-simple type can be drawn from the paper by M.Berger,[B]. The existence of the pseudo-Riemannian symmetric spaces of solvable type has been noticed already by E.Cartan,[C]. In the recent years, M.Cahen, M.Parker and N.Wallach have made a progress in the corresponding classification problems. The full classification is now known for the spaces of signature $(n,1)$, i.e., for the Lorentzian manifolds ([CW]). For the spaces of signature $(n,2)$ with the solvable transvection group the complete classification is given in [CP1]. (The structure of the spaces of mixed type is also described in a satisfactory way.) Finally, the complete classification of indecomposable pseudo-Riemannian symmetric spaces of mixed type (and of arbitrary signature) is known for the dimensions $n < 8$ ([CP2]).

# APPENDICES

## Appendix A. A digest of the theory of connections.

Let us recall briefly some well-known concepts and formulas from the theory of connections. (Cf.[KN I].) "Differentiable" always means "of class $C^\infty$".

Let $M$ be a differentiable manifold, $G$ a Lie group and $P(M,G)$ a principal fibre bundle over $M$. Thus $G$ acts on $P$ freely to the right and this action is simply transitive on each fibre of $P$.

For $u \in P$, let $G_u$ denote the subspace of $T_u(P)$ consisting of vectors which are tangent to the fibre through $u$. A connection $\Gamma$ in the bundle $P(M,G)$ is an assignement of a subspace $Q_u$ of $T_u(P)$ to each $u \in P$ such that

(a) $T_u(P) = G_u + Q_u$ (direct sum of vector spaces)

(b) $Q_{ua} = (R_a)_* Q_u$ for every $u \in P$ and $a \in G$ where $R_a$ is the transformation of $P$ induced by $a \in G$, $R_a u = ua$,

(c) $Q_u$ depends differentiably on $u$.

For each $u \in P$, $G_u$ is called the vertical subspace of $T_u(P)$, and $Q_u$ is called the horizontal subspace of $T_u(P)$ with respect to the connection $\Gamma$. Each tangent vector $X \in T_u(P)$ can be uniquely written as $X = Y + Z$, where $Y \in G_u$ and $Z \in Q_u$.

If $\alpha$ is a vector-valued differential k-form on $P(M,G)$, then the exterior covariant differential $D\alpha$ (with respect to $\Gamma$) is defined by the formula

$$(D\alpha)(X_1,\dots,X_{k+1}) = (d\alpha)(hX_1,\dots,hX_{k+1})$$

where $hX_i$ denote the horizontal components of $X_i$.

Given a connection $\Gamma$ in $P$, we define a 1-form $\omega$ on $P$ with values in the Lie algebra $\underline{g}$ of $G$ as follows: for each $X \in T_u(P)$ take the vertical component $Y$ of $X$ and define $\omega(X)$ as the unique element $A$ of the Lie algebra $\underline{g}$ such that $(d(u\cdot \exp tA)/dt)_{t=0} = Y$. Now, $\omega(X) = 0$ if and only if $X$ is horizontal. The form $\omega$ is called the connection form of the given connection $\Gamma$.

In particular, the set of all tangent frames to a manifold $M$ is a principal fibre bundle $L(M,GL(n,R))$, called the principal frame bundle of $M$.

An affine connection on $M$ is a bilinear map $\nabla\colon \mathfrak{X}(M)\times \mathfrak{X}(M)\longrightarrow \mathfrak{X}(M)$ written in the form $(X,Y)\longrightarrow \nabla_X Y$, satisfying the following axioms:

(i) $\nabla_{fX}Y = f(\nabla_X Y)$

(ii) $\nabla_X fY = (Xf)Y + f(\nabla_X Y)$ $\qquad X, Y \in \mathfrak{X}(M), \quad f \in \mathcal{F}(M).$

It is well-known that, for a point $p \in M$, the value $(\nabla_X Y)_p$ depends only on the value $X_p$ and on the germ of $Y$ at $p$. Thus, for any vector $u \in T_p(M)$ and any local vector field $Y$ in a neighborhood of $p$ we can define the vector $\nabla_u Y \in T_p(M)$, The vector $\nabla_u Y$ is called the covariant derivative of $Y$ with respect to $u$.

There is a bijective correspondence between the connections in the principal frame bundle $L(M) = L(M,GL(n,R))$ and the affine connections on $M$. Therefore, we often identify the corresponding objects.

Let now $G \subset GL(n,R)$ be a Lie subgroup and $P(M,G) \subset L(M,GL(n,R))$ a principal subbundle, i.e., a G-structure on $M$. Each connection in $P(M,G)$ can be extended in a unique way to a connection in $L(M)$.

Let $\Gamma$ be a connection in $P$, $\omega$ its connection form and $\Omega = D\omega$ the exterior covariant differential of $\omega$. $\Omega$ is called the curvature form of the connection $\Gamma$. We have the first structural equation

$$(A1) \qquad d\omega(X,Y) = -\tfrac{1}{2}[\omega(X),\omega(Y)] + \Omega(X,Y) \qquad X, Y \in T_u(P), \quad u \in P,$$

(with values in the Lie algebra $\underline{g} \subset \underline{gl}(n,R)$).

Further, let us recall that the canonical form $\theta$ of $P$ (independently of $\Gamma$) is the $R^n$-valued 1-form on $P$ defined by $\theta(X) = u^{-1}\cdot\pi_*(X)$ for $u \in P$, $X \in T_u(P)$. Here $\pi: P \longrightarrow M$ denotes the bundle projection, and each $u \in P$ is considered as a map of $R^n$ onto $T_{\pi(u)}(M)$. The torsion form $\Theta$ of $\Gamma$ on $P$ is defined as the exterior covariant differential $D\theta$ of $\theta$ with respect to $\Gamma$. It is uniquely determined by the second structural equation

$$(A2) \qquad d\theta(X,Y) = -\tfrac{1}{2}[\omega(X)\cdot\theta(Y) - \omega(Y)\cdot\theta(X)] + \Theta(X,Y),$$

for $X, Y \in T_u(P)$, $u \in P$ (with values in $R^n$).

Consider the torsion and curvature tensor fields of the corresponding affine connection on $M$:

$$(A3) \qquad T(X,Y) = \nabla_X Y - \nabla_Y X - [X,Y]$$

$$(A4) \qquad R(X,Y)Z = [\nabla_X,\nabla_Y]Z - \nabla_{[X,Y]}Z \qquad X,Y,Z \in \mathfrak{X}(M)$$

Then $T$ and $R$ are also defined by the following formulas by means of $\Theta$ and $\Omega$ respectively:

( A5 ) $T(X,Y) = u(2\Theta(\tilde{X},\tilde{Y}))$ for $X, Y \in T_x(M)$

( A6 ) $R(X,Y)Z = u(2\Omega(\tilde{X},\tilde{Y})(u^{-1}Z))$ for $X, Y, Z \in T_x(M)$,

where $u$ is any element of $P$ such that $\pi(u) = x$, and $\tilde{X}, \tilde{Y} \in T_u(P)$ are arbitrary lifts of $X, Y$ respectively.

Remark. Considering $u\cdot(2\Omega(\tilde{X},\tilde{Y}))$ as a "singular frame" at $x$, we can also write instead of (A6):

( A7 ) $R(X,Y)Z = u\cdot(2\Omega(\tilde{X},\tilde{Y}))(u^{-1}Z)$.

For the torsion tensor field $T$ and the curvature tensor field $R$ of an affine connection $\nabla$ on $M$ we have the well-known Bianchi identities:

( A8 ) $\mathfrak{S}(R(X,Y)Z = \mathfrak{S}\{T(T(X,Y),Z) + (\nabla_X T)(Y,Z)\}$

( A9 ) $\mathfrak{S}\{(\nabla_Z R)(X,Y) + R(T(X,Y),Z)\} = 0$.

(Here $\mathfrak{S}$ denotes the cyclic sum with respect to $X, Y, Z$.)

The following result is known from the theory of connections: Let $\Gamma$ be a connection in $L(M)$, $\nabla$ the corresponding affine connection on the manifold $M$, and $f: M \longrightarrow M$ a diffeomorphism. Then the connection $\Gamma$ is invariant with respect to the induced automorphism $\tilde{f}: L(M) \longrightarrow L(M)$ of the frame bundle if and only if $\nabla$ is invariant with respect to $f$ in the sense that

( A10 ) $f_*(\nabla_X Y) = \nabla_{f_* X} f_* Y$ for every $X, Y \in \mathfrak{X}(M)$.

$f$ is then called an affine transformation of the affine manifold $(M,\nabla)$.

More generally, if $(M,\nabla)$, $(M',\nabla')$ are two manifolds with the affine connections, then a diffeomorphism $f: M \longrightarrow M'$ is called an affine map if (A10) holds with the symbol $\nabla'$ on the right-hand side.

Let $\Gamma$ be a fixed connection in $L(M)$. Let $I \subset R$ be an arbitrary interval. A differentiable curve $\alpha: I \longrightarrow L(M)$ is said to be horizontal if all the tangent vectors $(d\alpha/dt)$ are horizontal with respect to $\Gamma$. To any differentiable curve $\gamma: I \longrightarrow M$ and any pair $(t_o,u_o)$ where $t_o \in I$ and $u_o \in \pi^{-1}(\gamma(t_o))$ there is a unique horizontal curve $\tilde{\gamma}: I \longrightarrow L(M)$ such that $\gamma = \pi\circ\tilde{\gamma}$ and $\tilde{\gamma}(t_o) = u_o$. If $I = \langle a,b\rangle$ is a closed finite interval, then we can assign in this way a unique frame $u_1 \in \pi^{-1}(\gamma(b))$ to any frame $u_o \in \pi^{-1}(\gamma(a))$. Hence we get an isomorphism $h_\gamma: T_{\gamma(a)}M \longrightarrow T_{\gamma(b)}M$ which is independent of the choice of $u_o \in \pi^{-1}(\gamma(a))$, and it is called the parallel transport along $\gamma$.

A differentiable map $v(t): I \longrightarrow T(M)$ is called a vector field along $\gamma: I \longrightarrow M$ if $v(t) \in T_{\gamma(t)}M$ for each $t \in I$. Such a vector field $v(t)$ is said to be parallel if, for any subinterval $\langle t_1, t_2 \rangle \subset I$, the vector $v(t_2)$ is the parallel translate of $v(t_1)$ along the arc $\gamma|\langle t_1, t_2 \rangle$.

A curve $\gamma: I \longrightarrow M$ is called a geodesic if the tangent vector field $v(t) = \frac{d\gamma}{dt}$ along $\gamma$ is parallel. It is well-known that, for each point $p \in M$ and each tangent vector $X \in T_p(M)$, there is a unique maximal geodesic $\gamma_X: I \longrightarrow M$ $(I \ni 0)$ with the properties $\gamma_X(0) = p$, $(d\gamma_X(0)/dt) = X$. The affine connection $\nabla$ corresponding to $\Gamma$ is said to be complete if all maximal geodesics $\gamma_X$, $X \in T(M)$, are defined on the whole real line $R$.

Let $\gamma: I \longrightarrow M$ be a regular differentiable curve. Then the parallelism along $\gamma$ can be expressed in terms of the affine connection $\nabla$ as follows: a vector field $v(t)$ along $\gamma$ is parallel if and only if $\nabla_{\frac{d\gamma}{dt}}(v(t)) = 0$ for all $t \in I$. Here $\nabla_{\frac{d\gamma}{dt}} v(t)$ is uniquely defined, for each $t \in I$, as the covariant derivative $\nabla_{\frac{d\gamma}{dt}} W$, where $W$ is an arbitrary (differentiable) local vector field in the neighborhood of $\gamma(t)$ such that $v(\tau) = W_{\gamma(\tau)}$ for all $\tau$ near $t$.

For every differentiable vector field $X$ on $M$, we have defined the operator $\nabla_X: Y \longrightarrow \nabla_X Y$ on vector fields from $\mathfrak{X}(M)$. There is a unique extension of this operator to the algebra $\mathcal{T}(M)$ of all differentiable tensor fields on $M$ such that the following holds:

(i) $\nabla_X$ is a derivation of $\mathcal{T}(M)$ with respect to the tensor product,

(ii) $\nabla_X$ preserves the type of any tensor field,

(iii) $\nabla_X$ commutes with the contractions,

(iv) $\nabla_X Y$ has the usual meaning for any vector field $Y$, and $\nabla_X f = Xf$ for any differentiable function.

Let $(M, \nabla)$ be a manifold with an affine connection. To any point $p \in M$ there is a neighborhood $N_o$ of the null vector in $T_p(M)$ such that the geodesics $\gamma_X$ are defined at the value $t = 1$, for all $X \in N_o$. Define the exponential map $\mathrm{Exp}_p: N_o \longrightarrow M$ by the formula $\mathrm{Exp}_p(X) = \gamma_X(1)$, $X \in N_o$. The image $\mathrm{Exp}_p(N_o) = N_p$ is called a normal neighborhood of $p$ if the map $Exp_p$ is a diffeomorphism of $N_o$ on $N_p$. A normal neighborhood always exists to each point $p \in M$.

The normal coordinates in a normal neighborhood are defined by the map $(x_1,\dots,x_n) \mapsto Exp_p(x_1e_1+\dots+x_ne_n)$ where $\{e_1,\dots,e_n\}$ is a basis in $T_p(M)$.

Let $\gamma: \langle a,b \rangle \longrightarrow M$ be a closed piece-wise differentiable curve, i.e. such that $\gamma(a) = \gamma(b) = p$. We can define the parallel transport along the whole of $\gamma$ using the parallel transports along the differentiable pieces of $\gamma$. All transformations $h_\gamma: T_p(M) \longrightarrow T_p(M)$ arising in this way form a Lie group $\Psi(p)$, called the holonomy group of M with reference point $p$. If $M$ is connected, then all holonomy groups $\Psi(x)$, $x \in M$, are mutually isomorphic. Choosing a frame $u_o$ at $p \in M$, we can define the holonomy group $\Phi(u_o) \subset GL(n,R)$ with reference frame $u_o$; here $\Phi(u_o) \cong \Psi(p)$. Any two holonomy groups $\Phi(u_o)$, $\Phi(u_1)$, $(u_o, u_1 \in L(M))$ are conjugate to each other in $GL(n,R)$.

## Appendix B. Some theorems from differential geometry.

Affine maps, isometries, holonomy. (Cf. [KN I] for more details.)

B1). Let $M$, $M'$ be manifolds with affine connections (or Riemannian manifolds) and let $f,g$, be affine maps (or isometries, respectively) of $M$ into $M'$. If $M$ is connected and $(f_*)_x = (g_*)_x$ for some $x \in M$, then $f$ and $g$ coincide on $M$.

B2). Let $M$ be a connected, simply connected analytic manifold with an analytic affine connection. Let $M'$ be an analytic manifold with a complete analytic affine connection. Then every affine map $f_U$ of a connected open subset $U$ of $M$ into $M'$ can be uniquely extended to an affine map $f$ of $M$ into $M'$.

B3). Let $M$ and $M'$ be analytic Riemannian manifolds. If $M$ is connected and simply connected and if $M'$ is complete, then every isometric immersion $f_U$ of a connected open subset $U$ of $M$ into $M'$ can be uniquely extended to an isometric immersion $f$ of $M$ into $M'$.

B4). Let $M$ and $M'$ be connected and simply connected, complete analytic Riemannian manifolds. Then every isometry between connected open subsets of $M$ and $M'$ can be uniquely extended to an isometry between $M$ and $M'$.

B5). Let $M$ be a connected differentiable manifold with an affine connection, and let $\Psi(x)$ be the holonomy group of $M$ with reference point $x$. Then the Lie algebra $\underline{g}(x)$ of $\Psi(x)$ is equal to the subspace of $End(T_x(M))$ spanned by all endomorphisms of the form $(\tau R)(X,Y) = \tau^{-1} \circ R(\tau X, \tau Y) \circ \tau$, where $X, Y \in T_x(M)$ and $\tau$ is the parallel transport along an arbitrary piecewise differentiable curve starting from $x$.

Affine manifolds with parallel curvature and parallel torsion.

B6). Let $M$ be a differentiable manifold with an affine connection such that $\nabla T = \nabla R = 0$. With respect to any atlas consisting of normal coordinate systems, $M$ is an analytic manifold and the connection $\nabla$ is also analytic.

B7). Let $M$ and $M'$ be differentiable manifolds with the affine connections $\nabla$ and $\nabla'$ respectively. Assume $\nabla T = \nabla R = 0$, $\nabla' T' = \nabla' R' = 0$. If $F$ is a linear isomorphism of $T_{x_0}(M)$ onto $T_{y_0}(M')$ and maps the tensors $T_{x_0}$ and $R_{x_0}$ at $x_0$ into the tensors $T'_{y_0}$ and $R'_{y_0}$ at $y_0$ respectively, then there is an affine isomorphism $f$ of a normal neighborhood $U_{x_0}$ onto a normal neighborhood $V_{y_0}$ such that $f(x_0) = y_0$ and $f_{*x_0} = F$. Explicitly, we have $f = Exp_{y_0} \circ F \circ Exp_{x_0}^{-1}$.

B8). Let $M$ and $M'$ in B7) be connected, simply connected and complete. Then there exists a unique affine isomorphism $f$ of $M$ onto $M'$ such that $f(x_0) = y_0$ and the differential of $f$ at $x_0$ coincides with $F$.

B9). Let $M$ be a differentiable manifold with an affine connection such that $\nabla T = \nabla R = 0$. Then the Lie algebra $\underline{g}(x)$ of the holonomy group $\Psi(x)$ is spanned by all endomorphisms of the form $R(X,Y)$, $X, Y \in T_x(M)$.

Remark. The proof of B9) follows immediately from B5) and from the property $\tau R = R$.

# R E F E R E N C E S


[B] M.Berger: Les espaces symétriques non compacts. Ann.Sci. Ecole Norm.Sup. 74 (1957), 85-177.

[Bo1] M.Božek: Existence of generalized symmetric Riemannian spaces with solvable isometry group. To appear in Čas.Pěst.Mat. (Praha).

[Bo2] - " - : Existence of generalized affine symmetric spaces of arbitrary order. To appear in Colloquium Math. (Warsaw).

[Bo3] - " - : Ph.D. Thesis, Praha 1976.

[Ca] E.Cartan: Sur une classe remarquable d'espaces de Riemann. Bull.Soc.Math. France 54 (1926), 214-264.

[Ch] C.Chevalley: Theory of Lie Groups I. Princeton Univ.Press, 1946.

[CP1] M.Cahen, M.Parker: Sur des classes d'espaces pseudo-riemanniens symétriques. Bull.Soc.Math.Belg. 22 (1970), No 4, 339-354.

[CP2] - " - , - " - : Espaces pseudo-riemanniens symétriques. Preprint, Université Libre de Bruxelles.

[CW] M.Cahen, N.Wallach: Lorentzian symmetric spaces. Bull.Am.Math.Soc. 76 (1970), 585-591.

[F1] A.S.Fedenko: Homogeneous $\varphi$-spaces and spaces with symmetries. Vestnik BGU (Minsk), Serie I, 1972, No 2, 25-30 (Russian).

[F2] - " - : Regular spaces with symmetries. Matem.zametki, AN SSSR, 14 (1973), No 1, 113-120 (Russian).

[F3] - " - : Spaces with symmetries (Russian, a booklet). Izd. BGU Minsk, 1977.

[GL1] P.J.Graham, A.J.Ledger: Sur une classe de s-variétés riemanniennes ou affines. C.R.Acad.Sci. Paris 267 (1967), 947-948.

[GL2] - " - , - " - : s-Regular Manifolds. Differential Geometry - in honour of Kentaro Yano, Tokyo 1972, 133-144.

[Gr] A.Gray: Riemannian manifolds with geodesic symmetries of order 3. J.Differential Geometry 7 (1972), No 3-4, 343-369.

[H] S.Helgason: Differential Geometry and Symmetric Spaces. Pure and Appl.Math. No 12, 1962, Academic Press, New York and London.

[H*] S.Helgason: Differential Geometry, Lie Groups, and Symmetric spaces,
Pure and Appl.Math., 1978,
Academic Press, New York - San Francisco - London.

[J] N.Jacobson: Lie Algebras.
Pure and Appl.Math. No 10,
Interscience Publ., John Wiley and Sons, New York - London.

[K1] O.Kowalski: Riemannian manifolds with general symmetries.
Math. Z. 136 (1974), No 2, 137-150.

[K2] - " - : Generalized symmetric spaces
(Preliminary communication).
Comm.Math.Univ. Carolinae, 15, 2 (1974), 361-375.

[K3] - " - : Classification of generalized symmetric
Riemannian spaces of dimension $n \leq 5$.
Rozpravy ČSAV, Řada MPV, No 8, 85 (1975).

[K4] - " - : Generalized pointwise symmetric spaces.
Comm.Math.Univ. Carolinae, 16, 3 (1975), 459-467.

[K5] - " - : Existence of generalized symmetric Riemannian
spaces of arbitrary order.
J.Differential Geometry 12 (1977), No 2, 203-208.

[K6 - " - : Generalized symmetric Riemannian spaces.
Period.Math.Hung. 8 (2), (1977), 181-184.

[K7] - " - : Smooth and affine s-manifolds.
Period.Math.Hung. 8 (3-4), 1977, 299-311.

[K8] - " - : Generalized affine symmetric spaces.
Math.Nachr. 80 (1977), 205-208.

[K9] - " - : On unitary automorphisms of solvable Lie algebras.
Pacific J.Math. 82 (1979), No 1, 133-143.

[K10] - " - : Affine reductive spaces.
Beiträge zur Algebra und Geometrie 8 (1979), 99-105.

[K11] - " - : Free periodic isometries of Riemannian manifolds,
J.London Math.Soc. (to appear).

[Ka] V.G.Kac: Automorphisms of finite order of semisimple
Lie algebras,
Funkcional. Analiz i Priloženija 3 (1969), 94-96.

[KL] O.Kowalski, A.J.Ledger: Regular s-structures on manifolds,
preprint, Liverpool 1972.

[KNI] [KNII] S.Kobayashi, K.Nomizu: Foundations of Differential Geometry.
Interscience Publ., New York - London,
Vol.I (1963), Vol.II (1969).

[L1] A.J.Ledger: Espaces de Riemann symétriques géneralisés.
C.R.Acad.Sci. Paris 264 (1967), 947-948.

[L2] - " - : s-Manifolds.
Chronik der 4.Panhellen. Math. Versammlung, Patras 1971, 91-93.

[LO] A.J.Ledger, M.Obata: Affine and Riemannian s-manifolds. J.Differential Geometry 2 (1968), No 4, 451-459.

[LP1] A.J.Ledger, B.Pettit: Compact quadratic s-manifolds. Comment.Math.Helv. 51 (1976), 105-131.

[LP2] - " - , - " - : Classification of metrizable regular s-manifolds with integrable symmetry tensor field. J.Math.Soc. Japan 28 (1976), No 4, 668-675.

[Lo1] O.Loos: Symmetric spaces, Vol.I: General Theory. Math. Lecture Note Series, W.A.Benjamin Inc., New York - Amsterdam 1969.

[Lo2] - " - : An intrinsic characterization of fibre bundles associated with homogeneous spaces defined by Lie group automorphisms. Abh.Math.Sem.Univ. Hamburg 37 (1972), 160-179.

[N] K.Nomizu: Invariant affine connections on homogeneous spaces. Amer.J.Math. 76 (1954) No 1, 33-65.

[S1] N.A.Stepanov: On reductivity of factor-spaces generated by endomorphisms of Lie groups, Izv. VUZ, Matematika, 1967, No 2, 74-79 (Russian).

[S2] - " - : Basic facts of the theory of $\varphi$-spaces. Izv. VUZ, Matematika, 1967, No 3, 88-95 (Russian).

[S3] - " - : Homogeneous 3-cyclic spaces. Izv. VUZ, Matematika, 1967, No 12, 65-74 (Russian).

[TL1] Gr.Tsagas, A.J.Ledger: Classification of simply connected four-dimensional RR-manifolds. Trans.Amer.Math.Soc. 219 (1976), 189-210.

[TL2] - " - , - " - : Riemannian s-manifolds. Preprint (Univ. of Patras, Univ. of Liverpool).

[VF] V.I.Vedernikov, A.S.Fedenko: Symmetric spaces and their generalizations. A survey article in: Algebra, Topology, Geometry, Vol. 14 M., VINITI AN SSSR, 1976 (Russian).

[W1] S.Wegrzynowski: Representation of generalized affine symmetric spaces by s-structures. Demonstratio Math. IX (1976), No 4, 1-21.

[W2] - " - : Classification of generalized affine symmetric spaces of dimension $n \leq 4$. To appear in Dissertationes Math. (Warsaw), 1980.

[Wi] D.J.Winter: On groups of automorphisms of Lie algebras. J.Alg. 8 (1968), 131-142.

[Wo] J.A.Wolf: Spaces of constant curvature. New York, Mc Graw-Hill, 1967.

[WoG] J.A.Wolf, A.Gray: Homogeneous spaces defined by Lie group automorphisms. J.Differential Geometry 2 (1968), No 1-2, 77-159.

# SUBJECT INDEX


admissible s-structure 111
affine connection 173
    complete 176
affine
    locally symmetric space 45
    map 175
    reductive space 41
    symmetric space 45
    s-structure 168
    transformation 175
amalgamation (of s-manifolds) 99
amalgamating decomposition 99
automorphism of a regular s-manifold 47
autoparallel submanifold 89

Bianchi identities 33, 175

canonical connection
    of a local regular s-structure 25, 70
    of a reductive homogeneous space 29
    of a regular s-manifold 47, 52
canonical form on the principal frame bundle 174
Cartan (-) connection 41
centerless infinitesimal s-manifold 100
characteristic variety (in the theory of eigenvalues) 116
closed subset of eigenvalues 108
complexification
    of a regular s-manifold 103
    of a generalized symmetric Riemannian space 105
connection form 173
covariant derivative 174
curvature form 174
curvature tensor field 174

derivation (of a manifold with multiplication) 48
direct product of regular s-manifolds 93
direct sum of infinitesimal s-manifolds 94

elementary transvection 57
exterior covariant differential 173

Fitting Lemma 53
Fitting 0-component, 1-component 53
foliations (in regular s-manifolds, in generalized symmetric spaces) 91, 98
free point (of a Riemannian manifold) 128

generalized affine symmetric space 111
  of semi-simple type 114
  of solvable type 114
  primitive 115
  unitary 113
generalized pointwise symmetric R. space 164
generalized symmetric Hermitian space 102
generalized symmetric Riemannian space 8
  proper 134
  without infinitesimal rotations 104
group of automorphisms of a regular s-manifold 47
group of transvections
  of an affine manifold 36
  of a regular s-manifold 58

Hermitian regular s-structure (s-manifold) 102
holonomy group 177

induced G-invariant metric 11
infinitesimal
  isomorphism of regular s-manifolds 74
  model of a (locally) regular s-manifold 74, 82
  s-manifold 73, 82
integrable tensor field S 88
invariant
  almost complex structure 84
  almost Hermitian structure 86
  complex structure 86
  foliation 91
  infinitesimal submanifold 90
  Kählerian structure 87
  submanifold 90
  tensor field 84

irreducible
generalized affine symmetric space 161
infinitesimal s-manifold 94
regular s-manifold 93
Riemannian space 20
set of eigenvalues 109
isomorphism
of infinitesimal s-manifolds 74, 82
of regular s-manifolds 74, 81

k-symmetric Riemannian manifold 8
Killing form 11

lattice point (in the theory of eigenvalues) 121
Levi decomposition 97
local
automorphism 69
geodesic symmetry 1, 45
isomorphism 69
regular s-structure 23, 68, 81
symmetry 24, 69
local regular s-triplet 76
effective 77
prime 77
locally regular s-manifold 69
Riemannian 81
Lorentzian symmetric space 172

metrizable regular s-manifold 65

nearly Kähler structure 171
Nijenhuis tensor 86, 88

order
of a free point 128
of a generalized affine symmetric space 113
of a generalized symmetric Riemannian space 8
of a periodic tensor structure 128
of an s-structure 4

parallel transport 176
periodic tensor structure 128

pointed
    affine manifold 41
    regular s-manifold 60
prime
    reductive homogeneous space 41
    regular homogeneous s-manifold 53
primitive generalized affine symmetric space 115
principal frame bundle 173
proper generalized symmetric R. space 134
pseudoduality 106, 107
pseudo-Riemannian symmetric space 172

reducible
    infinitesimal s-manifold 94
    regular s-manifold 93
    set of eigenvalues 109
Reduction Theorem (in the theory of connections) 36
reductive homogeneous space 10, 27
regular s-manifold 47
    of finite order 62
    of semi-simple type 96
    of solvable type 96
regular homogeneous s-manifold 53
regular s-structure 74, 81
    parallel, non-parallel 22, 23, 168
Riemannian space
    generalized symmetric 8
    homogeneous 9
    k-symmetric 8
    locally symmetric 1
    symmetric 1
    without infinitesimal rotations 65, 104

semi-simple symmetries 112
set of eigenvalues of a regular s-manifold 108
similarity (of locally regular s-manifolds) 69
s-structure 4
    of finite order 4, 113
    of infinite order 113
    of order k 4
    regular 7, 74, 81
    Riemannian 4

structural equations (of a connection) 174
symmetry
    affine 45
    geodesic 1, 45
    generalized 2, 47
    in a distributive groupoid 46, 47
symmetric space (by O.Loos) 45
system of eigenvalues (of a regular s-structure) 118

$\theta$-variety (in the theory of eigenvalues) 116
torsion form 174
torsion tensor field 174

weakly invariant
    subspace 92
    submanifold 92
    tensor field 88

# NOTATION INDEX

(See also the list of standard denotations.)


| | |
|---|---|
| $s_p$ | 1 |
| $S_p$ | 1 |
| $(M,g)$ | 1 |
| $\{s_x : x \in M\}$ | 2 |
| $Cl(\{s_x\})$ | 2 |
| $Cl(s_p)$ | 4 |
| $I(M,p)$ | 4 |
| $I(M_p)$ | 4 |
| $O(M_p)$ | 5 |
| $Cl(S_p)$ | 5 |
| $S$ | 6 |
| $Cl(\{s_x\},p)$ | 7 |
| $G^{\sigma}$, $(G^{\sigma})^{\bullet}$ | 9, 53 |
| $K_{\ell}$ | 41 |
| $Tr(M,\nabla)$ | 36 |
| $Tr(M)$ | 36, 58 |
| $Tr^{*}(M)$ | 39 |
| $(M,\mu)$ | 47 |
| $Aut(M)$ | 47 |
| $Der(M)$ | 49 |
| $x^{-1}\cdot y$ | 49 |
| $L(v)$ | 50 |
| $(M,\{s_x\})$ | 52 |
| $(G,H,\sigma)$ | 53 |
| $\underline{g}_{0A}$, $\underline{g}_{1A}$ | 53 |
| $Tr(M,\{s_x\})$ | 58 |
| $Cl^{a}(\{s_x\})$ | 65 |
| $\mathcal{A}^n$ | 117 |
| $\mathcal{B}^n$ | 119 |
| $\mathcal{C}^n$ | 119 |
| $\mathcal{D}^n$ | 120 |
| $\Lambda_n$ | 121, 123 |
| $\mathcal{L}_n$ | 121 |
| $\omega$ | 173 |
| $\theta$ | 174 |
| $\Omega$ | 174 |
| $\Theta$ | 174 |

Vol. 640: J. L. Dupont, Curvature and Characteristic Classes. X, 175 pages. 1978.

Vol. 641: Séminaire d'Algèbre Paul Dubreil, Proceedings Paris 1976–1977. Edité par M. P. Malliavin. IV, 367 pages. 1978.

Vol. 642: Theory and Applications of Graphs, Proceedings, Michigan 1976. Edited by Y. Alavi and D. R. Lick. XIV, 635 pages. 1978.

Vol. 643: M. Davis, Multiaxial Actions on Manifolds. VI, 141 pages. 1978.

Vol. 644: Vector Space Measures and Applications I, Proceedings 1977. Edited by R. M. Aron and S. Dineen. VIII, 451 pages. 1978.

Vol. 645: Vector Space Measures and Applications II, Proceedings 1977. Edited by R. M. Aron and S. Dineen. VIII, 218 pages. 1978.

Vol. 646: O. Tammi, Extremum Problems for Bounded Univalent Functions. VIII, 313 pages. 1978.

Vol. 647: L. J. Ratliff, Jr., Chain Conjectures in Ring Theory. VIII, 133 pages. 1978.

Vol. 648: Nonlinear Partial Differential Equations and Applications, Proceedings, Indiana 1976–1977. Edited by J. M. Chadam. VI, 206 pages. 1978.

Vol. 649: Séminaire de Probabilités XII, Proceedings, Strasbourg, 1976–1977. Edité par C. Dellacherie, P. A. Meyer et M. Weil. VIII, 805 pages. 1978.

Vol. 650: C*-Algebras and Applications to Physics. Proceedings 1977. Edited by H. Araki and R. V. Kadison. V, 192 pages. 1978.

Vol. 651: P. W. Michor, Functors and Categories of Banach Spaces. VI, 99 pages. 1978.

Vol. 652: Differential Topology, Foliations and Gelfand-Fuks-Cohomology, Proceedings 1976. Edited by P. A. Schweitzer. XIV, 252 pages. 1978.

Vol. 653: Locally Interacting Systems and Their Application in Biology. Proceedings, 1976. Edited by R. L. Dobrushin, V. I. Kryukov and A. L. Toom. XI, 202 pages. 1978.

Vol. 654: J. P. Buhler, Icosahedral Golois Representations. III, 143 pages. 1978.

Vol. 655: R. Baeza, Quadratic Forms Over Semilocal Rings. VI, 199 pages. 1978.

Vol. 656: Probability Theory on Vector Spaces. Proceedings, 1977. Edited by A. Weron. VIII, 274 pages. 1978.

Vol. 657: Geometric Applications of Homotopy Theory I, Proceedings 1977. Edited by M. G. Barratt and M. E. Mahowald. VIII, 459 pages. 1978.

Vol. 658: Geometric Applications of Homotopy Theory II, Proceedings 1977. Edited by M. G. Barratt and M. E. Mahowald. VIII, 487 pages. 1978.

Vol. 659: Bruckner, Differentiation of Real Functions. X, 247 pages. 1978.

Vol. 660: Equations aux Dérivée Partielles. Proceedings, 1977. Edité par Pham The Lai. VI, 216 pages. 1978.

Vol. 661: P. T. Johnstone, R. Paré, R. D. Rosebrugh, D. Schumacher, R. J. Wood, and G. C. Wraith, Indexed Categories and Their Applications. VII, 260 pages. 1978.

Vol. 662: Akin, The Metric Theory of Banach Manifolds. XIX, 306 pages. 1978.

Vol. 663: J. F. Berglund, H. D. Junghenn, P. Milnes, Compact Right Topological Semigroups and Generalizations of Almost Periodicity. X, 243 pages. 1978.

Vol. 664: Algebraic and Geometric Topology, Proceedings, 1977. Edited by K. C. Millett. XI, 240 pages. 1978.

Vol. 665: Journées d'Analyse Non Linéaire. Proceedings, 1977. Edité par P. Bénilan et J. Robert. VIII, 256 pages. 1978.

Vol. 666: B. Beauzamy, Espaces d'Interpolation Réels: Topologie et Géometrie. X, 104 pages. 1978.

Vol. 667: J. Gilewicz, Approximants de Padé. XIV, 511 pages. 1978.

Vol. 668: The Structure of Attractors in Dynamical Systems. Proceedings, 1977. Edited by J. C. Martin, N. G. Markley and W. Perrizo. VI, 264 pages. 1978.

Vol. 669: Higher Set Theory. Proceedings, 1977. Edited by G. H. Müller and D. S. Scott. XII, 476 pages. 1978.

Vol. 670: Fonctions de Plusieurs Variables Complexes III, Proceedings, 1977. Edité par F. Norguet. XII, 394 pages. 1978.

Vol. 671: R. T. Smythe and J. C. Wierman, First-Passage Perculation on the Square Lattice. VIII, 196 pages. 1978.

Vol. 672: R. L. Taylor, Stochastic Convergence of Weighted Sums of Random Elements in Linear Spaces. VII, 216 pages. 1978.

Vol. 673: Algebraic Topology, Proceedings 1977. Edited by P. Hoffman, R. Piccinini and D. Sjerve. VI, 278 pages. 1978.

Vol. 674: Z. Fiedorowicz and S. Priddy, Homology of Classical Groups Over Finite Fields and Their Associated Infinite Loop Spaces. VI, 434 pages. 1978.

Vol. 675: J. Galambos and S. Kotz, Characterizations of Probability Distributions. VIII, 169 pages. 1978.

Vol. 676: Differential Geometrical Methods in Mathematical Physics II, Proceedings, 1977. Edited by K. Bleuler, H. R. Petry and A. Reetz. VI, 626 pages. 1978.

Vol. 677: Séminaire Bourbaki, vol. 1976/77, Exposés 489–506. IV, 264 pages. 1978.

Vol. 678: D. Dacunha-Castelle, H. Heyer et B. Roynette. Ecole d'Eté de Probabilités de Saint-Flour. VII-1977. Edité par P. L. Hennequin. IX, 379 pages. 1978.

Vol. 679: Numerical Treatment of Differential Equations in Applications, Proceedings, 1977. Edited by R. Ansorge and W. Törnig. IX, 163 pages. 1978.

Vol. 680: Mathematical Control Theory, Proceedings, 1977. Edited by W. A. Coppel. IX, 257 pages. 1978.

Vol. 681: Séminaire de Théorie du Potentiel Paris, No. 3, Directeurs: M. Brelot, G. Choquet et J. Deny. Rédacteurs: F. Hirsch et G. Mokobodzki. VII, 294 pages. 1978.

Vol. 682: G. D. James, The Representation Theory of the Symmetric Groups. V, 156 pages. 1978.

Vol. 683: Variétés Analytiques Compactes, Proceedings, 1977. Edité par Y. Hervier et A. Hirschowitz. V, 248 pages. 1978.

Vol. 684: E. E. Rosinger, Distributions and Nonlinear Partial Differential Equations. XI, 146 pages. 1978.

Vol. 685: Knot Theory, Proceedings, 1977. Edited by J. C. Hausmann. VII, 311 pages. 1978.

Vol. 686: Combinatorial Mathematics, Proceedings, 1977. Edited by D. A. Holton and J. Seberry. IX, 353 pages. 1978.

Vol. 687: Algebraic Geometry, Proceedings, 1977. Edited by L. D. Olson. V, 244 pages. 1978.

Vol. 688: J. Dydak and J. Segal, Shape Theory. VI, 150 pages. 1978.

Vol. 689: Cabal Seminar 76–77, Proceedings, 1976–77. Edited by A.S. Kechris and Y. N. Moschovakis. V, 282 pages. 1978.

Vol. 690: W. J. J. Rey, Robust Statistical Methods. VI, 128 pages. 1978.

Vol. 691: G. Viennot, Algèbres de Lie Libres et Monoïdes Libres. III, 124 pages. 1978.

Vol. 692: T. Husain and S. M. Khaleelulla, Barrelledness in Topological and Ordered Vector Spaces. IX, 258 pages. 1978.

Vol. 693: Hilbert Space Operators, Proceedings, 1977. Edited by J. M. Bachar Jr. and D. W. Hadwin. VIII, 184 pages. 1978.

Vol. 694: Séminaire Pierre Lelong – Henri Skoda (Analyse) Année 1976/77. VII, 334 pages. 1978.

Vol. 695: Measure Theory Applications to Stochastic Analysis, Proceedings, 1977. Edited by G. Kallianpur and D. Kölzow. XII, 261 pages. 1978.

Vol. 696: P. J. Feinsilver, Special Functions, Probability Semigroups, and Hamiltonian Flows. VI, 112 pages. 1978.

Vol. 697: Topics in Algebra, Proceedings, 1978. Edited by M. F. Newman. XI, 229 pages. 1978.

Vol. 698: E. Grosswald, Bessel Polynomials. XIV, 182 pages. 1978.

Vol. 699: R. E. Greene and H.-H. Wu, Function Theory on Manifolds Which Possess a Pole. III, 215 pages. 1979.

Vol. 700: Module Theory, Proceedings, 1977. Edited by C. Faith and S. Wiegand. X, 239 pages. 1979.

Vol. 701: Functional Analysis Methods in Numerical Analysis, Proceedings, 1977. Edited by M. Zuhair Nashed. VII, 333 pages. 1979.

Vol. 702: Yuri N. Bibikov, Local Theory of Nonlinear Analytic Ordinary Differential Equations. IX, 147 pages. 1979.

Vol. 703: Equadiff IV, Proceedings, 1977. Edited by J. Fábera. XIX, 441 pages. 1979.

Vol. 704: Computing Methods in Applied Sciences and Engineering, 1977, I. Proceedings, 1977. Edited by R. Glowinski and J. L. Lions. VI, 391 pages. 1979.

Vol. 705: O. Forster und K. Knorr, Konstruktion verseller Familien kompakter komplexer Räume. VII, 141 Seiten. 1979.

Vol. 706: Probability Measures on Groups, Proceedings, 1978. Edited by H. Heyer. XIII, 348 pages. 1979.

Vol. 707: R. Zielke, Discontinuous Čebyšev Systems. VI, 111 pages. 1979.

Vol. 708: J. P. Jouanolou, Equations de Pfaff algébriques. V, 255 pages. 1979.

Vol. 709: Probability in Banach Spaces II. Proceedings, 1978. Edited by A. Beck. V, 205 pages. 1979.

Vol. 710: Séminaire Bourbaki vol. 1977/78, Exposés 507–524. IV, 328 pages. 1979.

Vol. 711: Asymptotic Analysis. Edited by F. Verhulst. V, 240 pages. 1979.

Vol. 712: Equations Différentielles et Systèmes de Pfaff dans le Champ Complexe. Edité par R. Gérard et J.-P. Ramis. V, 364 pages. 1979.

Vol. 713: Séminaire de Théorie du Potentiel, Paris No. 4. Edité par F. Hirsch et G. Mokobodzki. VII, 281 pages. 1979.

Vol. 714: J. Jacod, Calcul Stochastique et Problèmes de Martingales. X, 539 pages. 1979.

Vol. 715: Inder Bir S. Passi, Group Rings and Their Augmentation Ideals. VI, 137 pages. 1979.

Vol. 716: M. A. Scheunert, The Theory of Lie Superalgebras. X, 271 pages. 1979.

Vol. 717: Grosser, Bidualräume und Vervollständigungen von Banachmoduln. III, 209 pages. 1979.

Vol. 718: J. Ferrante and C. W. Rackoff, The Computational Complexity of Logical Theories. X, 243 pages. 1979.

Vol. 719: Categorial Topology, Proceedings, 1978. Edited by H. Herrlich and G. Preuß. XII, 420 pages. 1979.

Vol. 720: E. Dubinsky, The Structure of Nuclear Fréchet Spaces. V, 187 pages. 1979.

Vol. 721: Séminaire de Probabilités XIII. Proceedings, Strasbourg, 1977/78. Edité par C. Dellacherie, P. A. Meyer et M. Weil. VII, 647 pages. 1979.

Vol. 722: Topology of Low-Dimensional Manifolds. Proceedings, 1977. Edited by R. Fenn. VI, 154 pages. 1979.

Vol. 723: W. Brandal, Commutative Rings whose Finitely Generated Modules Decompose. II, 116 pages. 1979.

Vol. 724: D. Griffeath, Additive and Cancellative Interacting Particle Systems. V, 108 pages. 1979.

Vol. 725: Algèbres d'Opérateurs. Proceedings, 1978. Edité par P. de la Harpe. VII, 309 pages. 1979.

Vol. 726: Y.-C. Wong, Schwartz Spaces, Nuclear Spaces and Tensor Products. VI, 418 pages. 1979.

Vol. 727: Y. Saito, Spectral Representations for Schrödinger Operators With Long-Range Potentials. V, 149 pages. 1979.

Vol. 728: Non-Commutative Harmonic Analysis. Proceedings, 1978. Edited by J. Carmona and M. Vergne. V, 244 pages. 1979.

Vol. 729: Ergodic Theory. Proceedings, 1978. Edited by M. Denker and K. Jacobs. XII, 209 pages. 1979.

Vol. 730: Functional Differential Equations and Approximation of Fixed Points. Proceedings, 1978. Edited by H.-O. Peitgen and H.-O. Walther. XV, 503 pages. 1979.

Vol. 731: Y. Nakagami and M. Takesaki, Duality for Crossed Products of von Neumann Algebras. IX, 139 pages. 1979.

Vol. 732: Algebraic Geometry. Proceedings, 1978. Edited by K. Lønsted. IV, 658 pages. 1979.

Vol. 733: F. Bloom, Modern Differential Geometric Techniques in the Theory of Continuous Distributions of Dislocations. XII, 206 pages. 1979.

Vol. 734: Ring Theory, Waterloo, 1978. Proceedings, 1978. Edited by D. Handelman and J. Lawrence. XI, 352 pages. 1979.

Vol. 735: B. Aupetit, Propriétés Spectrales des Algèbres de Banach. XII, 192 pages. 1979.

Vol. 736: E. Behrends, M-Structure and the Banach-Stone Theorem. X, 217 pages. 1979.

Vol. 737: Volterra Equations. Proceedings 1978. Edited by S.-O. Londen and O. J. Staffans. VIII, 314 pages. 1979.

Vol. 738: P. E. Conner, Differentiable Periodic Maps. 2nd edition, IV, 181 pages. 1979.

Vol. 739: Analyse Harmonique sur les Groupes de Lie II. Proceedings, 1976–78. Edited by P. Eymard et al. VI, 646 pages. 1979.

Vol. 740: Séminaire d'Algèbre Paul Dubreil. Proceedings, 1977–7[illegible] Edited by M.-P. Malliavin. V, 456 pages. 1979.

Vol. 741: Algebraic Topology, Waterloo 1978. Proceedings. Edited by P. Hoffman and V. Snaith. XI, 655 pages. 1979.

Vol. 742: K. Clancey, Seminormal Operators. VII, 125 pages. 1979.

Vol. 743: Romanian-Finnish Seminar on Complex Analysis. Proceedings, 1976. Edited by C. Andreian Cazacu et al. XVI, 713 pages. 1979.

Vol. 744: I. Reiner and K. W. Roggenkamp, Integral Representations. VIII, 275 pages. 1979.

Vol. 745: D. K. Haley, Equational Compactness in Rings. III, 16[illegible] pages. 1979.

Vol. 746: P. Hoffman, τ-Rings and Wreath Product Representations. V, 148 pages. 1979.

Vol. 747: Complex Analysis, Joensuu 1978. Proceedings, 1978. Edited by I. Laine, O. Lehto and T. Sorvali. XV, 450 pages. 1979.

Vol. 748: Combinatorial Mathematics VI. Proceedings, 1978. Edited by A. F. Horadam and W. D. Wallis. IX, 206 pages. 1979.

Vol. 749: V. Girault and P.-A. Raviart, Finite Element Approximation of the Navier-Stokes Equations. VII, 200 pages. 1979.

Vol. 750: J. C. Jantzen, Moduln mit einem höchsten Gewicht. [illegible] 195 Seiten. 1979.

Vol. 751: Number Theory, Carbondale 1979. Proceedings. Edited by M. B. Nathanson. V, 342 pages. 1979.

Vol. 752: M. Barr, *-Autonomous Categories. VI, 140 pages. 197[illegible]

Vol. 753: Applications of Sheaves. Proceedings, 1977. Edited by M. Fourman, C. Mulvey and D. Scott. XIV, 779 pages. 1979.

Vol. 754: O. A. Laudal, Formal Moduli of Algebraic Structures. I[illegible] 161 pages. 1979.

Vol. 755: Global Analysis. Proceedings, 1978. Edited by M. Grme[illegible] and J. E. Marsden. VII, 377 pages. 1979.

Vol. 756: H. O. Cordes, Elliptic Pseudo-Differential Operators – A Abstract Theory. IX, 331 pages. 1979.

Vol. 757: Smoothing Techniques for Curve Estimation. Proceedings, 1979. Edited by Th. Gasser and M. Rosenblatt. V, 245 pages. 197[illegible]

Vol. 758: C. Năstăsescu and F. Van Oystaeyen; Graded and Filtered Rings and Modules. X, 148 pages. 1979.